Photography and War

exposures

EXPOSURES is a series of books on photography designed to explore the rich history of the medium from thematic perspectives. Each title presents a striking collection of images and an engaging, accessible text that offers intriguing insights into a specific theme or subject.

Series editors: Mark Haworth-Booth and Peter Hamilton

Also published

Photography and Africa Erin Haney
Photography and Anthropology Christopher Pinney
Photography and Archaeology Frederick N. Bohrer
Photography and Australia Helen Ennis
Photography and China Claire Roberts
Photography and Cinema David Campany
Photography and Death Audrey Linkman
Photography and Egypt Maria Golia
Photography and Exploration James R. Ryan
Photography and Flight Denis Cosgrove and William L. Fox
Photography and Germany Andrés Mario Zervigón
Photography and Humour Louis Kaplan
Photography and Ireland Justin Carville
Photography and Italy Maria Antonella Pelizzari
Photography and Japan Karen Fraser
Photography and Literature François Brunet
Photography and Science Kelley Wilder
Photography and Spirit John Harvey
Photography and Tibet Clare Harris
Photography and Travel Graham Smith
Photography and the USA Mick Gidley
Photography and War Pippa Oldfield

Photography and War

Pippa Oldfield

reaktion books

To my parents

Published by Reaktion Books Ltd
Unit 32, Waterside
44–48 Wharf Road
London N1 7UX, UK
www.reaktionbooks.co.uk

First published 2019

Printed and bound in China

A catalogue record for this book is available from the British Library

ISBN 978 1 78914 145 0

Contents

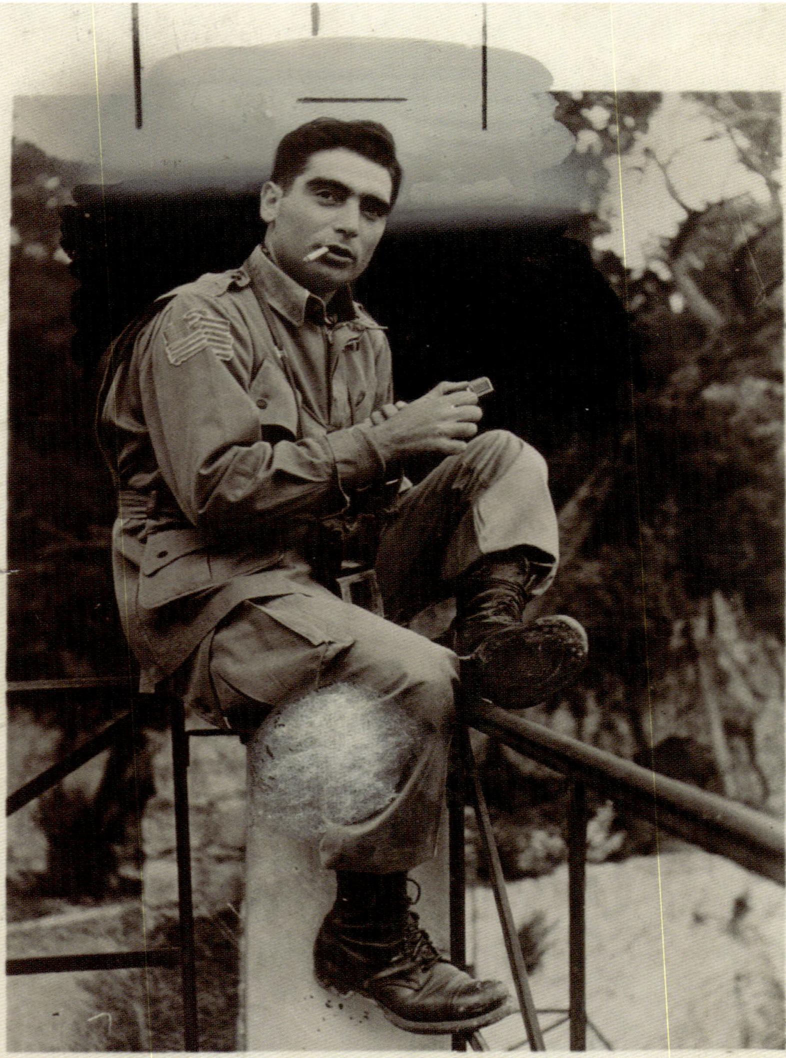

Robert Capa, author of SLIGHTLY OUT OF FOCUS, published by Henry Holt and Company

Introduction

The classic war photograph is a dramatic black-and-white image, made in the combat zone by a risk-taking professional photographer. Robert Capa, the pseudonym of a Hungarian émigré who made his name in the Spanish Civil War (1936–9), is one of the most enduring of these figures. His most famous photograph, apparently showing a Loyalist soldier at the moment of death by gunfire, is an icon of war photography and continues to generate discussion (see illus. 9).

'Classic' war photography, however, is only one manifestation of the intertwined pursuits of photography and war, the focus of this book. Innumerable photographic images, activities and technologies have been made in the context of war, in a bewildering array of ways. They may bear witness and share experiences, or be clandestine and hidden from sight. Some purport to offer truths; others invent fictions. They may be driven by the need to persuade and recruit, or to survey and spy. Some strive to evoke empathy, while others display contempt. They may be motivated by financial gain, or the wish to dominate and conquer. Still others aim to record, to remember and to reflect.

War is not simply an intermittent intrusion into everyday life, but is – for better or worse – embedded in societies and integral to their evolution. Consequently, many people have undertaken photographic work that responds to war in some way, without ever considering themselves bona fide war photographers. Photographic and military technologies, meanwhile, have advanced hand-in-hand, inflected by scientists and inventors, corporations and entrepreneurs, politicians and the armed forces. These multiple interactions between photography and war have resulted in many

1 George Rodger, Robert Capa in Naples, Italy, 1943, silver gelatin press print produced 1947.

collaborations, applications and consequences, sometimes unexpected. We might consider, for instance, the market for photographic postcards prompted by the Mexican Revolution (1910–20); the interpretation of photographic aerial reconnaissance to inform bomb targeting during the Second World War (1939–45); or the circulation of digital snapshots of prisoners abused by U.S. military personnel at Iraq's Abu Ghraib prison, made during the Iraq War (2003–11). Each of these disparate practices, though far removed from Robert Capa's 'classic' war photography, has something important to say about the prosecution, experience and impact of war.

Most visual anthologies of photography and war are presented chronologically. They tend to relate the histories of conflicts, sometimes alongside developments in photographic technologies, using images as illustrations.[1] Other books follow biographical or art-historical trajectories, presenting the exploits and visions of particular photographers in one or more conflicts.[2] While these approaches have produced many important and worthwhile publications, they also have their limitations. Photographs are often presented as unproblematic windows on the world that might reveal some kind of 'truth', particularly to the civilian viewer who has not experienced the horror of war at first hand. The focus is on certain aspects of war – the soldier's experience, military hardware and the drama of combat – at the expense of others that might be harder to photograph, less dramatic or even less 'glamorous'. As a result, many important aspects of war are overlooked, and the viewpoints of many people are ignored.

Photography and War aims to address these limitations, taking a thematic approach, expanding the range of material considered and recognizing that 'the truth of war' is a subjective and always shifting notion. Throughout this book, photographic products and processes are assumed to be highly mediated affairs, which are made, viewed and disseminated in specific circumstances and motivated by particular impulses. Close attention is paid to these contexts to help understand the work that photography might do in that particular instance.

The materiality of photographs – their qualities as physical objects in use – is often very revealing. Consider, for instance, the portrait of Robert Capa (illus. 1) made by fellow war correspondent George Rodger who, with

Capa and others, would co-found the collective Magnum Photos after the war. The photograph was taken in Naples in 1943, while Capa and Rodger were covering the Allied liberation of Italy. This particular print, however, was not made in 1943 but in 1947 when, according to the typewritten caption, the publishing company Henry Holt & Co. was promoting Capa's semi-fictionalized autobiography *Slightly Out of Focus*. Rodger is not credited as the photographer, suggesting that in this instance the subject of the image was more important than its creator. The background behind Capa's head has been painted out, presumably to produce a clearer image for halftone reproduction in a newspaper. Crop marks indicate that the image – already reduced from Rodger's original square format negative, most probably made with a Rolleiflex rather than a 35mm camera – was cropped further to create a tight headshot.[3] The photograph's surface is worn and grubby, befitting a press print that would have been loaned out to periodicals for a fee and then returned to the agency for subsequent distribution. All of these details suggest a complexity of function that goes beyond a straightforward representation of a famous war photographer. Rather, the materiality of the photograph bespeaks its function as a commodity circulating within a media economy, and Capa-the-war-photographer as a marketable 'product', a notion that is discussed in more detail in the following chapter.

Numerous sites play host to encounters with photography and war. Newspapers, magazines and books are perhaps the most obvious places to view war photographs, but there are many others. Images have been viewed in sophisticated metropolitan art galleries, and in rudimentary tents and cabins in military camps (illus. 2). They have been scrutinized in courts of law, and pasted into private photo albums. Images have been peered at through nineteenth-century stereoscopic viewers, and shared via twenty-first-century social media networks such as Twitter.

The encounter between viewer and image, and the nature of that encounter, is where we see photographs 'at work'. An early instance was vividly described by Oliver Wendell Holmes, a nineteenth-century cultural critic for the *New York Times*. In 1862 Holmes visited Mathew Brady's National Portrait Gallery in New York to view the exhibition *Incidents of the War*, featuring photographs made during the ongoing Civil War. 'Go into the Gallery when you may,' Holmes wrote, 'and you will see crowds gathering

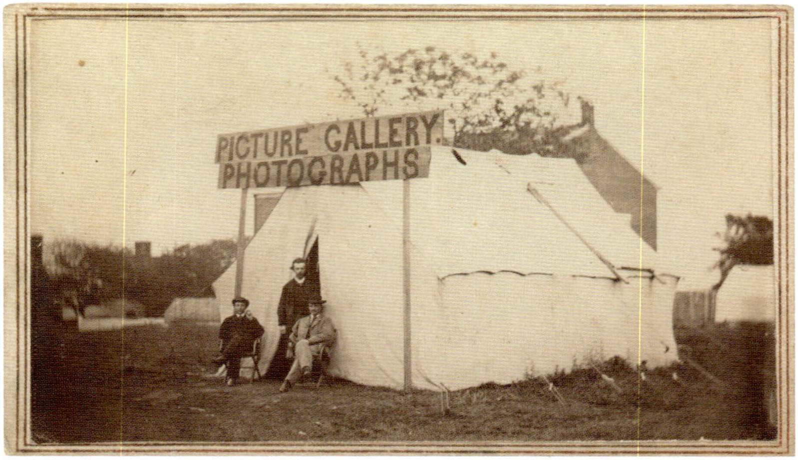

2 Itinerant commercial photographers outside tent serving as a gallery and studio during the American Civil War, 1861–5, albumen print on card.

around these pictures, some with tearful eyes, some with eyes that brim with pride, and all with swelling hearts.'[4] Some months later, Holmes returned to the gallery to view *The Dead of Antietam*, whose photographs depicted battlefields littered with the mangled corpses of soldiers. Holmes had a personal interest in the topic, having recently returned from Maryland in search of his son, a soldier who had been shot in the neck and – miraculously – had survived. Holmes climbed the stairs to the gallery to join the 'hushed, reverend groups standing around these weird copies of carnage'.[5] For many visitors, it was their first experience of photographs of war dead.

It is hard to imagine the shock this encounter provoked. Today, graphic images of violence are readily available, and are frequently encountered in print, on screen and via digital media. But in the mid-nineteenth century, photography was a recent invention, and photographic reproduction in newspapers was still impossible. People were accustomed to viewing war via illustrations, engravings and paintings, which tended to romanticize and sanitize the realities of armed conflict. Photographs, by contrast, seemed to collapse the distance between battlefield and civilian viewer. Holmes declared, 'Mr Brady has done something to bring home to us the terrible reality and earnestness of war. If he has not brought bodies and laid them on our dooryard and along the streets, he has done something very like it.'[6]

The notion of 'bringing home' the 'terrible reality' of war to a populace that might be far removed from the actual events has become a central tenet of war photography in its conventional photojournalistic mode – albeit one that has been much debated.

As far as is known, there are no photographs that depict the reactions of visitors to *Incidents of the War*. Nearly a hundred years later, however, the U.S. photojournalist Wayne Miller made a memorable image of onlookers to another New York exhibition, held at the Museum of Modern Art in 1955 (illus. 3).[7] The photographer's wife and young children stand in front of a large-scale print of an atomic bomb explosion, an image which recalls the Allied detonations at Hiroshima and Nagasaki in Japan at the close of the Second World War, ten years previously. The children turn to their mother in fear and incomprehension at the terrifying prospect of nuclear annihilation, a far cry from the 'swelling hearts' prompted by the American Civil War pictures described by Holmes.

The ostensible subject of Miller's image is the innocence of children in the atomic age and, perhaps, the hope that a new generation would not repeat the horrors wreaked upon Hiroshima and Nagasaki. Yet Miller's image conveys other aspects of the intertwined relationships between photography and war. There are military technologies at work here too. The photograph-within-a-photograph of the iconic 'mushroom cloud' was made by the U.S. military and supplied to the exhibition by the Federal Civil Defense Administration. The photograph's original function (as a record of the test *Mike*, a thermonuclear explosion from Operation Ivy at Enewetak Atoll in 1952) was to enable assessment and improvement of U.S. weaponry. The detonation of atomic bombs, moreover, was enabled by stereoscopic technology. This application of flash photography technology was developed by EG&G, the company co-founded by scientist Harold Edgerton, a photography pioneer more popularly known for his split-second depictions of milk drops and tennis players.[8] Viewed in this light, photography does far more than merely document war: it is actively deployed in the prosecution of warfare.

We might also consider the setting where Miller's wife and children contemplate the mushroom cloud: the *Family of Man* exhibition. This international blockbuster show was curated in the wake of the Second World War by Edward Steichen, director of the Department of Photography at the Museum of Modern Art, and the former director of the U.S. Naval Photographic Institute (and an expert in aerial photo-reconnaissance). The exhibition positioned photography as a kind of universal language that could bridge national divides, and convince viewers of the common bonds of humanity and the terrible futility of war. More cynically, *The Family of Man* was also a major exercise in cultural diplomacy at a time when U.S.–Soviet relations were deteriorating into the stalemate of the Cold War and a nuclear arms race. The exhibition was, therefore, a global branding exercise in all-American post-war values of consumerism, capitalist democracy and the nuclear family. Seen through these many lenses, Miller's multilayered image illustrates the intricacy and complexity of the relationships between photography and war.

To help untangle some of these relationships, this book is divided into five thematic chapters. Each brings together images and practices that are expressive of broader tendencies and cut across historical periods. The first chapter, 'Despatches from the Combat Zone', offers an account of war photography in its classic mode, considering the work of front-line photojournalists. Some of war's most famous photographers and iconic images are featured here: Roger Fenton's cannonball-strewn battlefield in the Crimea; Timothy O'Sullivan's corpses lying on the field after the Battle of Gettysburg; Yevgeny Khaldei's flag-raising at the Reichstag in Berlin. However, the chapter also pays attention to lesser-known conflicts, such as the War of the Triple Alliance against Paraguay (1865–70), and less illustrious photography, such as snapshots made by women ambulance drivers close to the trenches in the First World War. Issues of truth, authenticity and authority are often at stake, as well as the impact of photographs on those who view them.

The cultural figure of the 'truth-seeking', independent war photographer, however, has received disproportionate attention. War is, by and large, a state-sponsored exercise. Governments and their agencies, including the armed forces, have an intrinsic interest in controlling, promoting, justifying and defending their actions. Photographers have often found themselves enmeshed in the larger systems and structures for war. Motivated by patriotism or pragmatism, many photographers working in official or semi-official positions have done what was asked of them, supporting what might be termed 'the normalization' of war. At the same time, the business interests of photographic corporates have been forceful drivers of technologies and practices. The second chapter, 'Military Vision', sets the scene for these fundamental and often symbiotic relationships between photography and war. Drawing on technologies of reconnaissance, surveying and vision beyond the human eye, the discussion considers the connection between visual knowledge and victory through a wide range of techniques, from mapping and microfilm to aerial photography and infrared. The varied cases show that photography has advanced ever more sophisticated modes of warfare, while the needs of the military have, in turn, catalysed photographic innovation.

3 Wayne Miller, Miller's wife and children in front of a picture of a hydrogen bomb, at the exhibition *The Family of Man*, Museum of Modern Art, New York, 1963, silver gelatin print.

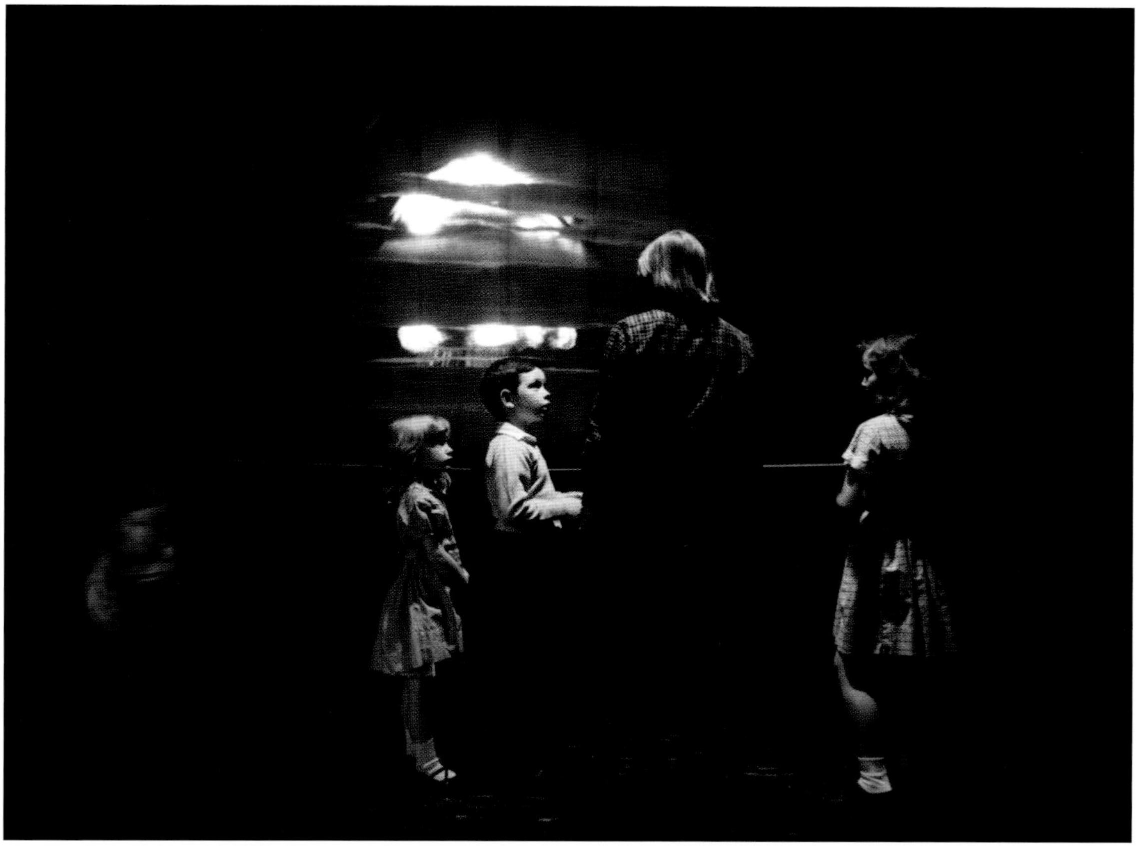

Chapter Three, 'Home Fronts', examines the experiences of non-combatants, as well as activities that are geographically far from official war zones. Yet, as the images reveal, war's operations often play out far closer to home than is conventionally assumed. Many images in this chapter are drawn from conflicts from the early twentieth century onwards, when the introduction of long-range weapons meant that civilian populations were within reach of military bombardment, and burgeoning notions of 'total war' and 'war effort' demanded active civilian involvement in war. The participation of women features strongly in this chapter, as do homeland operations such as propaganda and recruitment. Some of the key methods and approaches employed include photomontage, a technique particularly utilized by war protestors; civilian photography, including portraiture and snapshots; and persuasive print platforms such as magazines, posters and advertisements.

Photography's vital and persistent role in bearing witness is the subject of Chapter Four, 'Secrets and Exposures'. As demonstrated throughout this book, the assumption of photography as a 'naturally' truthful and objective medium is clearly problematic. And yet, the political value of photography's evidentiary power may, under specific circumstances, be made useful. Humanitarians, activists and legal crusaders have frequently used photographs as evidence to expose war crimes and make demands for truth and justice. Conversely, in the hands of perpetrators, photography has the capacity to dehumanize, humiliate and repress. The chapter considers the role of photography in conducting or indicting state terrorism, incarceration, torture and human rights abuses in numerous contexts, including the Holocaust of the Second World War (1939–45); the regime of the Khmer Rouge in Cambodia (1975–9); the so-called 'Dirty War' of Argentina's military dictatorship (1976–83); and more recently the Syrian Civil War (2011 and ongoing). While many of the images in this book are disturbing, viewers may find this chapter particularly harrowing.

The closing chapter, 'Legacies', focuses on the long-term consequences of war, considering how people are affected and how they remember, both as individuals and societies. While there are many ways to chart war's far-reaching trajectories, a particularly fruitful way has been 'aftermath photography', an approach that emerged at the end of the twentieth century and is gathering pace. Widely practised by contemporary artists and critically engaged documentary photographers, the aftermath approach investigates the traces of war, whether found in the landscape or on the human body, as well as war's psychological consequences. The works discussed demonstrate that the convenient binaries of 'war' and 'peace' are over-simplifications, and that, for many people, the impact of war continues to shape their lives. Finally, while the trauma of war may continue long after armed conflict has ended, its aftermath also holds the promise of healing, reconciliation and peacebuilding. While an in-depth investigation of the relationship between photography and peace would require far more space than is available here, this concluding chapter touches on the overlooked role that photography might play in prompting or recording activities that promote more peaceful societies.

As might be expected from this indication of content, I have defined 'war' in broad terms, to encompass armed conflict in many forms: revolution, 'cold' war, state terrorism, civil conflict, terrorism, state-sponsored violence and genocide, in addition to more conventionally accepted 'hot' wars and international conflicts between regular armies. The images I have chosen, moreover, have come from a huge range of sources: public and private, prestigious and humble. I have attempted to cast a wide net in terms of geographies and cultures, reflecting a more inclusive range of viewpoints than is usual for photographic anthologies of war. Recent scholarly work on the colonial, racial and gendered politics of photography makes it imperative to consider images by and for those often excluded from official histories.[9]

Photography was a Western invention, closely tied to the doctrines of modernization, industrialism, capitalism and colonialism. Photography beyond the West was initially a project of outsiders: of entrepreneurs, government-funded expeditions and military units, among other interested parties, and many of the early images featured in this book reflect these motivations. Nonetheless, a plethora of cultures and societies have used photographic technologies for their own ends, and a number of the images discussed hail from the Far East, the Indian subcontinent, Russia and the former Eastern Bloc, the Middle East and Latin America, as well as reflecting Black, indigenous, diasporic or multi-heritage viewpoints.

Special attention has been paid to the participation of women at the intersection of photography and war. Gender roles are thrown into sharp relief in wartime, with men stereotyped as soldiers, warriors and leaders, and women as mourners, non-combatants and victims. The zones of 'home' and 'front' have conventionally been seen as separate spheres, gendered as feminine/safe and masculine/dangerous respectively, despite the wealth of evidence that contests these neat binaries.[10] Women have historically been excluded from military roles and positioned as outsiders to war, and consequently granted less authority to comment than men. It is also true that, historically speaking, men have outnumbered women working in photography, and men still dominate the fields of 'hard' news, photojournalism and war photography. Yet, as this book shows, women have persistently been active participants in war, and repeatedly used photography to demonstrate their engagement. Female war photographers in the 'classic' photojournalistic mode, such as Gerda Taro and Lee Miller, have made important contributions to the genre. Moreover, as Val Williams has argued in her pathbreaking book *Warworks* (1994), women have also utilized photographic practices that go beyond the narrowly defined canon.

A reconsideration of war photography with attention to gender, therefore, is not a case of merely slotting female practitioners into the existing masculinist framework, but rather an opportunity to rethink what 'war photography' might be. Women have been active as inventors of technological processes; as military agents interpreting photographic surveillance; as snapshooters and amateurs; and as artists who have challenged the conventions of war photography. The more expansive approach taken in this book allows the substantial participation by women in photography and war to come into view. Moreover, the diversity of women's activities – and, by the same token, those of men – demonstrates that belligerence or pacifism may in no way be attributed to simple biological difference.

Given the limitations of space, I have chosen to omit many iconic images, especially where they have been ably discussed extensively elsewhere, such as Robert Capa's *D-Day Landings* (1945), Joe Rosenthal's *Flag Raising at Iwo Jima* (1945) and Larry Burrows's *Reaching Out* (1966).[11] Many of the images that I did include have rich and complex histories, and merit far more expansive critiques than there is space for here. There is a huge amount of literature devoted to, for instance, the photographs of Mathew Brady or the Abu Ghraib scandal, and I have signalled avenues for further research in the References and Select Bibliography. It would have been impossible, furthermore, to cover every conflict in the global history of war since the invention of photography, or unpick every aspect of the interwoven relationships between photography and war. Yet, while this book does not aim or claim to be exhaustive, I hope it may offer a succinct view into photography's perhaps most contested, complex and emotive subject and show how a consideration of photography may enrich our understanding of the experience and effects of war. Whether considered a brutal and repugnant practice, or a necessary – even desirable – aspect of politics and society, war remains a fundamental experience that shapes and transforms lives on a mass scale, from which photography is inseparable.

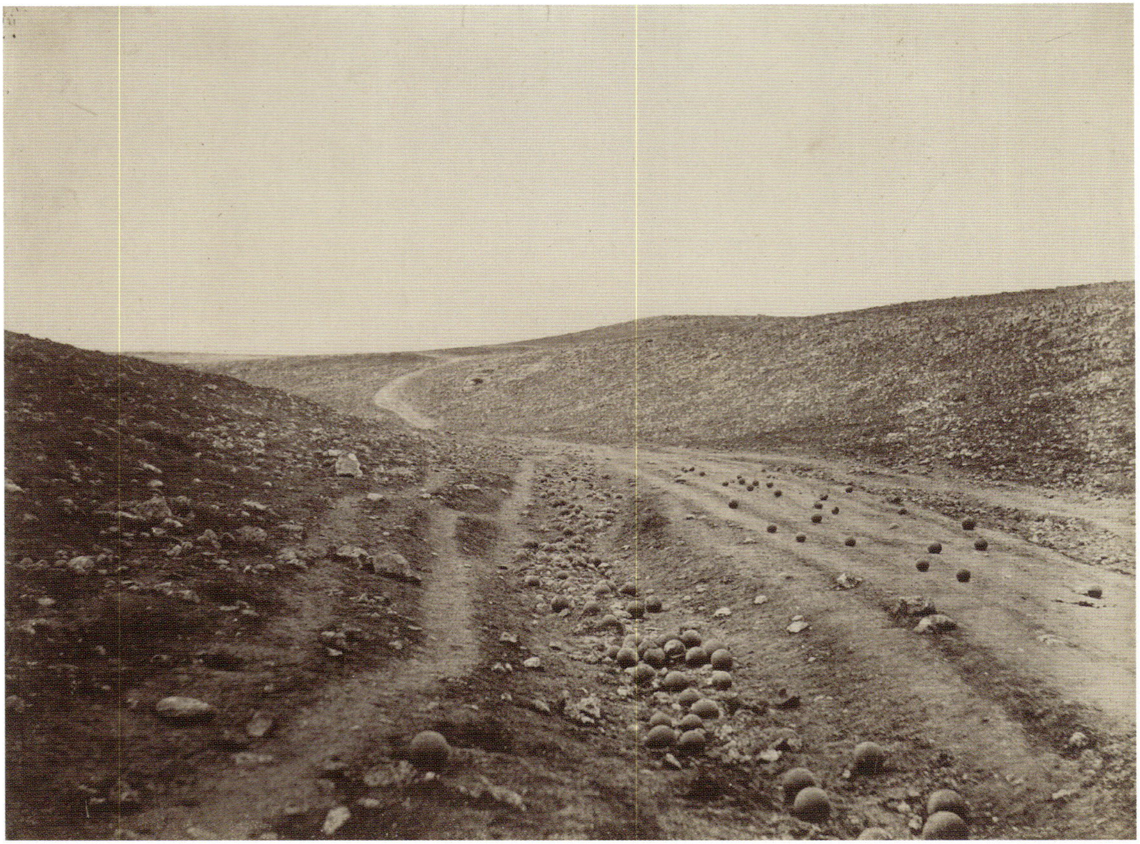

one

Despatches from the Combat Zone

In 1855 the British magazine *Punch* satirized the public appetite for photographs depicting the unfolding conflict in the Crimea. In a postscript to a letter to a soldier, a fictional female correspondent writes,

> P.S. I send you, dear Alfred, a complete Photographic apparatus, which will amuse you doubtlessly in your moments of leisure, and if you could send me home, dear, a good view of a nice battle, I should feel extremely obliged.
> P.S. No. 2. If you could take the view, dear, just in the moment of victory, I should like it all the better.[1]

While only a spoof, the passage foresees some of the themes and assumptions that would shape the nascent genre of war photography: eyewitness images made from the soldier's viewpoint in the heart of combat action; the use of photography to report from front line (characterized as military and masculine) to home front (civilian and feminine); and the promise of pleasurable aesthetic contemplation of scenes of war, viewed from a comfortable distance. These assumptions are examined – and challenged – in the pages that follow.

The genre of war photography tends to be dominated by an emphasis on the figure of the intrepid war photographer – typified by the great Robert Capa – who is motivated by adventure, the quest for 'truth', historical consciousness and humanitarian concerns. But this persona did not emerge unbidden, nor do war photographers, past or present, work in a vacuum. They are part of an economy that is supported by consumers and their desire

4 Roger Fenton, *Valley of the Shadow of Death*, 1855, salt print.

to view – like Alfred's sweetheart – events of the combat zone, a desire upon which various suppliers may capitalize.

The Crimean War (1853–6), an armed struggle between Russia and a Franco-British-Turkish alliance, was the first conflict to be extensively photographed and consumed through images.[2] Numerous photographers made records, including Carol Popp de Szathmari, Col. Jean-Charles Langlois, James Robertson and Felice Beato.[3] It was Roger Fenton, however, whose work set a precedent for conveying the sights of the far-off battlefield to audiences curious to see the events of war via a 'photographic apparatus'. Fenton, a well-to-do Englishman and vocal advocate for photography, was prompted in part by commercial motivations. He was commissioned and financed by British publisher Thomas Agnew & Sons, who had identified an opportunity to sell photographic prints to the public. Fenton also had the blessing of photography enthusiasts Queen Victoria and Prince Albert, as well as the cooperation of the British Army, which hoped Fenton's expedition might counter adverse press on the war.[4] Given Fenton's patrons, and his position as a prototypical 'embed' integrated with British troops, it is unsurprising that he produced dignified and sanitized images of conflict. Although Fenton encountered corpses in the aftermath of battle, he was under instructions not to jeopardize public support by suggesting that British troops were suffering.[5]

The logistical challenges of producing photographs in the field were considerable. The constraints of photographic technology – slow shutter speeds and limited light sensitivity of emulsions – meant that spontaneous images of action were impossible to capture. Instead, Fenton produced impressive staged portraits of soldiers in full regalia, who might have been posing in a smart London studio. Fenton's field views are, of necessity, stationary and detached rather than made in the heat of battle. His best-known image, *The Valley of the Shadow of Death* (1855), evokes Tennyson's emotive poem *The Charge of the Light Brigade*, written a year earlier to commemorate a daring but disastrous cavalry assault (illus. 4). For viewers accustomed to vivid paintings and illustrations of war, Fenton's unpopulated image must have seemed uncharacteristically bare and sombre. In contrast to the lithographs of William Simpson, an artist who was in the Crimea at the same time as Fenton, there are no dashing cavalry charges to be seen.[6] The only indications of combat are the spent cannon balls littering the valley, and even they were artfully redistributed by Fenton for greater dramatic effect.[7] The viewer must make a considerable imaginative leap to reconstruct the scenes of fighting from this barren and static image.

Nonetheless, the apparently unmediated realism of photography gave Fenton's images a distinctive power.[8] Owing to the impossibility of photo-mechanical reproduction in newspapers, the images were translated into wood engravings and were published in the popular periodical *Illustrated London News*.[9] Prints were displayed in at least 26 venues in the UK, and although the exhibition tour was a commercial failure in terms of sales, it received a reported two million visitors.[10] Fenton's *Valley of the Shadow of Death* remains an icon of war photography even today, foreshadowing the 'aftermath' approach of contemporary artists discussed in Chapter Five, 'Legacies'.

The American Civil War (1861–5) vastly expanded the market for viewing images of war. Photo-historians have invariably highlighted the remarkable contribution of Mathew Brady, a visionary entrepreneur who maintained portrait-studios and galleries in New York City and Washington, DC. Following his foray to the First Battle of Bull Run (Manassas) in 1861, Brady enlisted over twenty talented photographers, including George N. Barnard, Timothy O'Sullivan and Alexander Gardner, commissioning them to photograph in the field, and disseminated their images through his pioneering photo-agency.[11] Thanks to his talent for promotion – typified by the ubiquitous credit 'Photograph by Brady', regardless of who actually made the image – Brady became the best-known photographer of the war.

The economy of Civil War images, however, was far greater than a single name, or even a single agency. Countless other photographers adapted to new market conditions and transferred their operations to the combat zone, or at least close to it. Hardly had the dust settled on Fort Sumter before newspapers in the Confederate states were advertising 'Truly Superb Photographic Views', made 'From the Seat of War', which could be bought from book and periodical stores, studios and galleries.[12] 'War views' of this type encompassed battlefields, ports, ruins, encampments, troops, bridges and weaponry, some made by photographers whose names are now barely remembered.

As well as 'war views', there was a huge appetite for mass-produced portraits of famous military leaders, not to mention ambrotypes, tintypes

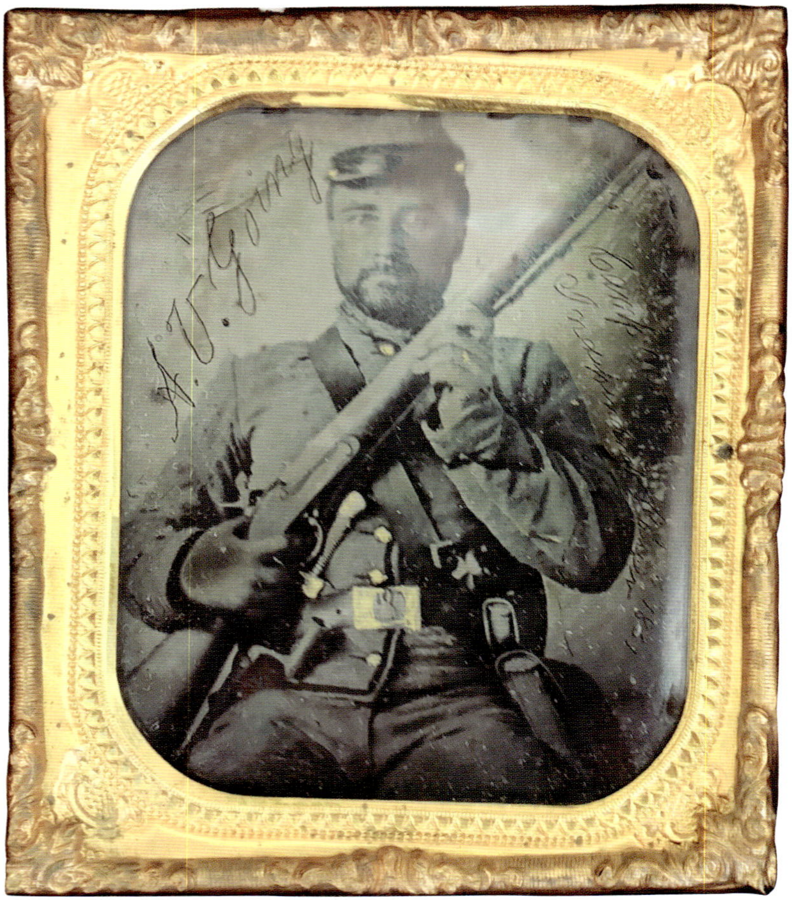

5 Elizabeth Beachbard, Amasa Vernon Going, 12th Louisiana Infantry Co. E, Camp Moore, Louisiana, 18 August 1861, ambrotype in case.

and *cartes de visite* of individual soldiers. Photographers capitalized on this demand, sometimes shutting up shop to follow military clientele as they went to war. The high-street ambrotypist Elizabeth Beachbard was one of the first to relocate, establishing a wooden cabin at Camp Moore, Louisiana, where she photographed Confederate soldiers prior to their deployment (illus. 5). By August 1862 the *New York Tribune* declared that no camp was free from the 'omnipresent artists in collodion and amber-bead varnish'.[13] Perhaps the most successful of these entrepreneurs were the Bergstresser Brothers of Pennsylvania, who were reputed to have made 160 one-dollar portraits of Union soldiers in one day alone, while based with the Army of the Potomac at Fredericksburg.[14]

As had been the case during the Crimean War, photo-mechanical reproduction in print media was still impossible. Attempts were made to produce photographically illustrated books by binding large original prints into a single album, such as Brady's *Incidents of the War* (1862) and *Gardner's Photographic Sketch Book of the War* (1866) (see illus. 13). Although these grand and handsome volumes subsequently received much acclaim, at the time they were commercial failures, produced in small numbers for an elite market that could afford to spend up to $150 on a luxury item.

During the conflict, people encountered images in other ways. Photographs were translated into wood engravings and published in popular illustrated periodicals such as *Harper's Weekly*, often appended with the caption 'from a photograph' as an indication of authenticity.[15] Original photographs were displayed everywhere from upmarket galleries such as Brady's on New York's Broadway (as discussed in the Introduction) to humble tents hastily erected near military encampments (see illus. 2). Photographs were sold in a range of sizes – as *cartes de visite*, small mounted album cards and various larger mounted prints – priced as low as 25 cents for the smallest.[16] An album card depicting the impressive battery at Yorktown, Virginia, is appended with a label verso declaring the view to be number 375

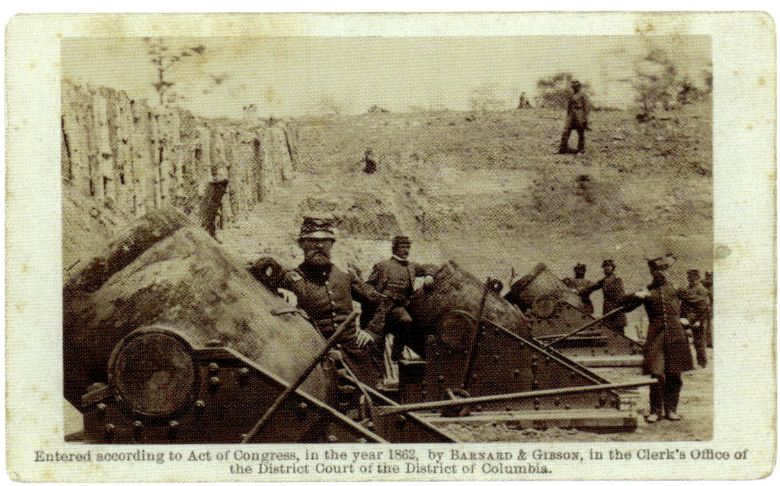

6 George N. Barnard and James F. Gibson, 'Battery, No. 4 – near Yorktown, mounting 10-13-inch mortars, each weighing 20,000 pounds. East-South end', 1862, albumen print on card, published by Brady Album Gallery.

in 'Brady's Album Gallery' (illus. 6). Purchasers were encouraged to collect and arrange the series in pre-slotted albums, for domestic and social display.[17] The copyright notices on both the front and reverse of the card highlight how the financial value of such enterprises was zealously guarded.

One of the most popular modes of dissemination, now virtually forgotten, was the stereograph. Developed as a means of rendering photographic scenes three-dimensional, stereographs comprised two almost-identical images taken with a twin-lens camera to mimic human binocular vision. The two prints were mounted on card and viewed through a special hand-held device, a stereoscope. The physical closeness demanded by the stereoscope, and the optical illusion of entering a three-dimensional world, offered an immersive, intimate experience that was distinct from viewing prints in an album or exhibition, and elicited something of a parlour craze. Manufacturers such as the New York-based E. & H. T. Anthony & Co. supplemented their existing stereographic inventory with war images supplied by Brady; Alexander Gardner produced and sold his own. Stereographs and other prints were available from shops, galleries and photo-studios as well as by mail order.[18] Gardner's *Catalogue of Photographic Incidents of the War* (1863), for example, provided listings of images in a bewildering array of sizes and options.[19] By the time of the Confederate surrender in May 1865, images of war had become far more commonplace than could ever have been imagined at the outset, and continued to circulate in the decades that followed.[20]

Photojournalism and the Cult of the War Photographer

There was clearly great potential for images to be reproduced in the printed press, if only the technology could be invented. In 1882 the German copper engraver George Meisenbach patented a revolutionary solution: the autotype or halftone process. By using a screen to translate photographs into dots of varying density, images could then be etched as plates for use in printing presses.[21] In 1880 the first photographic reproduction appeared in a newspaper, and by the end of the century, halftone technology was adopted by existing illustrated periodicals such as the *Illustrirte Zeitung* in Germany, the *Illustrated London News* in the UK and *L'Illustration* in France. Although photojournalism would not fully develop until the mid-1920s, photographic news in the press had begun.[22]

One of the first conflicts to be disseminated via halftone technology was the Spanish–American War of 1898, fought in Cuba and the Philippines. Eager periodicals despatched correspondents to Cuba in anticipation of hostilities, and the media tycoon William Randolph Hearst was even reputed to have told his staff, 'You furnish the pictures, and I'll furnish the war.'[23] Although apocryphal, the story exemplifies the growing influence attributed to illustrated print journalism, and the perception that war represented a commercial opportunity that could be exploited by the media. The U.S. periodical *Collier's* began cultivating a roster of photographic correspondents that included British émigré James (Jimmy) Hare, whose images boosted the periodical's circulation.[24] The Bain News Service, a photo agency established by George Grantham Bain in 1898, instructed Frances Benjamin Johnston to cover the story of the return of the U.S. fleet after its naval victory in the Philippines. A professional photographer with high-profile Washington connections, Johnston gained access to the USS *Olympia* anchored in Naples, where she photographed Admiral George Dewey, the 'Hero of Manila Bay'. She also made surprisingly intimate photographs of sailors showing the crew eating in their mess below-decks, reclining in their quarters and even dancing a male-only waltz (illus. 7). Female press photographers were still a minority, but Johnston was a vocal advocate for women's professionalization.[25]

By the turn of the century, a new cultural persona was emerging: the daredevil photographer who roved the world in search of conflicts, despatching images of violence and drama to be viewed in newspapers by a distant public. Some of these photographers began to achieve fame in their own right. Viktor Bulla, whose photographs of the Russo-Japanese War of 1904–5 appeared in the *Illustrated London News* and *Collier's*, was claimed to be the special correspondent of at least two Russian publications including *Iskry* and *Niva*. A portrait of the young photographer, reproduced in *Niva*, depicts him as a romantic figure attired in military uniform and bearskin Cossack hat, with camera cases slung across his chest.[26] At *Collier's*, meanwhile, Jimmy Hare was elevated from the status of unnamed 'Special Artist' to that of famed 'War Photographer', whose personal qualities of courage and determination were as much a part of the story as his pictures. In an early example of celebrity photographic endorsement, a Kodak advertisement published in *Collier's* emphasized Hare's credentials as a bona fide war photographer working 'under artillery fire' in the Russo-Japanese War.[27] A few years later, the *El Paso Daily Herald* set the seal on his star status, claiming that 'no war is official until it has been covered by Jimmy Hare's camera.'[28]

By the time of the Mexican Revolution of 1910, seasoned war photographers like Hare were joined by opportunist novices who converged on accessible sites of battle. Ciudad Juárez was described by the *New York Herald* in 1911 as 'lousy with freelance photographers', and the adjacent Texan city of El Paso even offered hotels for war tourists to view the battles between federal and rebel forces.[29] The Mexican reporter and photographer Agustín Casasola, whose pivotal role as an agent bears comparison with that of Mathew Brady some fifty years earlier, formed the Agencia Fotográfica Mexicana in 1912, which distributed images by Manuel Ramos, Hugo Brehme and countless others to periodicals and newspapers such as *Nueva Era* and *La*

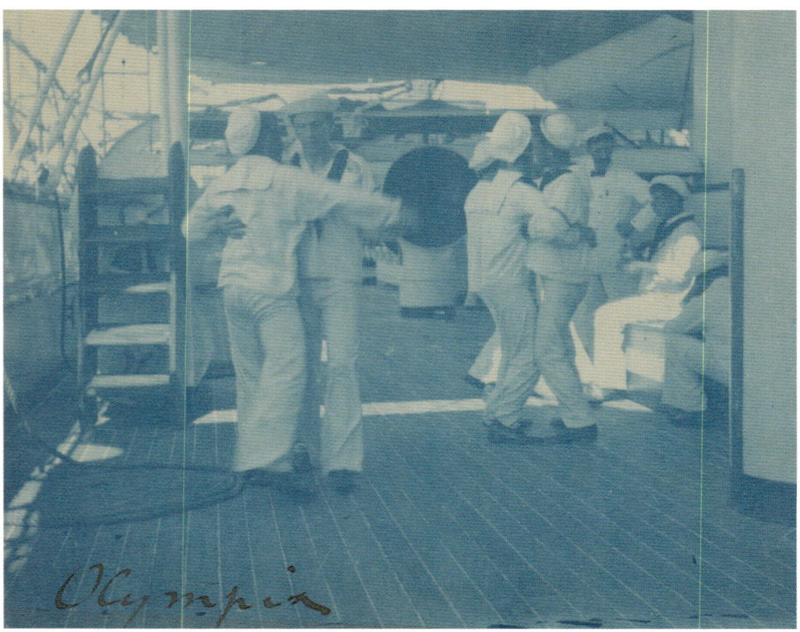

7 Francis Benjamin Johnston, *Olympia* (sailors waltzing onboard the USS *Olympia*), 1899, cyanotype.

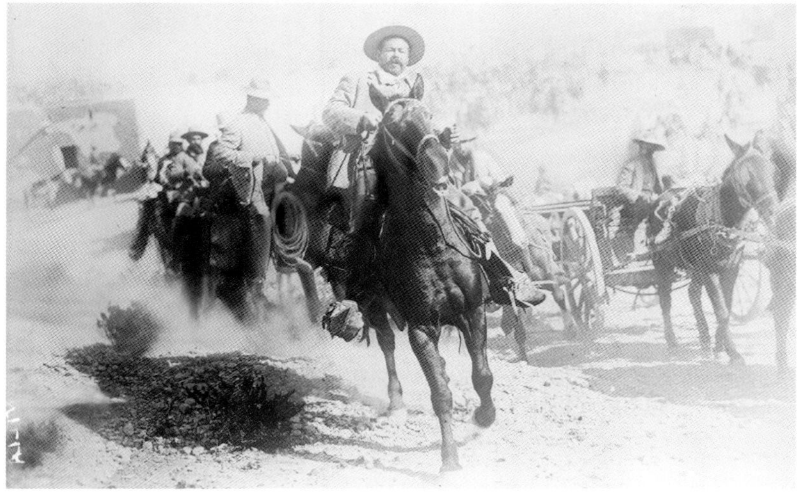

8 Mutual Film Company, Francisco 'Pancho' Villa at the Battle of Ojinga, Chihuahua, Mexico, 1914, image from black-and-white glass negative.

Ilustración Semanal.[30] The dissemination of images in illustrated journals was rivalled only by the photographic postcard industry, which counted large manufacturers such as U.S. manufacturer Walter H. Horne – at its peak in 1916 producing 5,000 cards a day – as well as numerous one-man-band operations. While postcards destined for U.S. markets stereotyped Mexicans as bandits and 'greasers', more nuanced portrayals came from Mexican photographers such as H. J. Gutiérrez and Jesús H. Abitia.[31]

Given the buoyancy of the market, it is no surprise that photographers and subjects sometimes colluded to fabricate profitable images. A striking image of the Mexican rebel leader 'Pancho' Villa on horseback in mid-gallop, purportedly made at the Battle of Ojinga, Chihuahua (1914), was distributed by the Bain News Service, and reproduced in the U.S. magazines *Leslie's* and *Collier's* (illus. 8). It is likely, however, that the image was actually made by a stills photographer working for the U.S. Mutual Film Corporation, to whom Villa had granted exclusive rights to film his battles in return for $25,000 to finance his military operations.[32] Villa, who welcomed photographers and reputedly fitted freight cars as rolling stock darkrooms, was sharply aware of the value of positive press.[33] The romantic horseback image attracted international sympathy for the charismatic general and his revolutionary cause, at least until relations with the U.S. military soured.

In the 1920s and '30s, photographic journalism in the mass media boomed. This growth was enabled by technological advances in reproduction (improved halftone and rotogravure processes); faster transmission of images (by 1935 'wire photo' services cut transmission time from days to hours); and faster film speeds and more portable cameras (the German Leica I, using 35mm film, was introduced in 1925). Photographically illustrated periodicals came into competition with a new wave of modern mass-market magazines: *Münchner Illustrierte Presse* (1923) and *Arbeiter-Illustrierte-Zeitung* (1925) in Germany; *Rotográfico* (*c.* 1926) in Mexico; *Vu* (1928) and *Regards* (1932) in France; *Life* (1936) and *Look* (1937) in the U.S.; and *Picture Post* in the UK (1938), among others.[34] These periodicals demanded eye-catching imagery, and dramatic news and documentary photography became key components.

Images of war were thus valuable commodities. In 1935, as Italy was poised to attack Ethiopia, *Time* editor Henry Luce was urged to hurry the launch of *Life*, on the grounds that 'a war, any sort of war, is going to be natural promotion for a picture magazine.'[35] In the event, it was the Spanish Civil War that would deliver on this promise. The popular struggle against Fascism was an accessible, even 'glamorous' war, sparking international interest. Photographers converged on Spain, including the Hungarians Kati Horna and Robert Capa (Endre Friedmann), the German Gerda Taro (Gerta Pohorylle), the Austrian-Australian Margaret Michaelis and Chim (David Seymour) from Poland, as well as native photographers such as the Catalan Agustí Centelles. The greatest star to emerge from this melting pot of war photography was undoubtedly the young and talented Capa, hailed by *Picture Post* in 1938 as 'The Greatest War-Photographer in the World'.[36] Reportedly declaring that 'if your pictures aren't good enough, you're not close enough', Capa swept the viewer into the heart of the action.[37] His iconic image of the death of a Loyalist soldier (1936) takes proximity to its logical conclusion, apparently capturing the very moment a soldier is killed by a bullet, just metres away from the photographer (illus. 9). The image was reproduced in a multitude of mass-market periodicals, with *Vu* first off the mark on 2 September 1936 with a photo-essay devoted to Capa's reportage, and also featured on the cover of Capa's 1938 book *Death in the Making*. The claim, much later, that the image was most likely staged has not diminished Capa's

9 'Robert Capa's camera catches a Spanish soldier the instant he is dropped by a bullet through the head in front of Cordoba', *Life* magazine, 12 July 1937.

LIFE

Vol. 3, No. 2

JULY 12, 1937

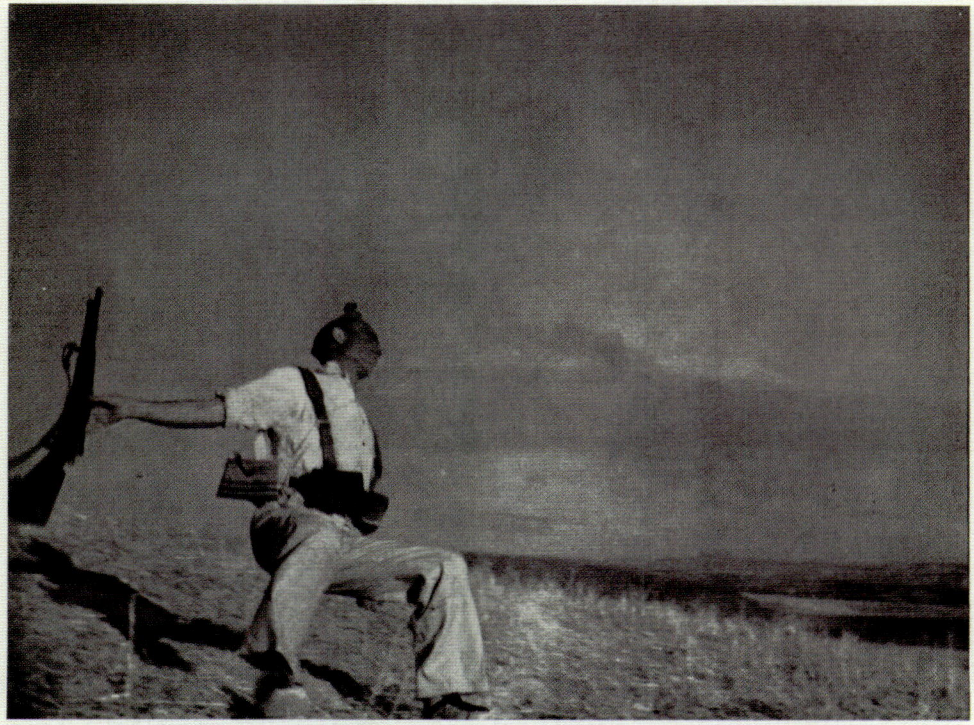

ROBERT CAPA'S CAMERA CATCHES A SPANISH SOLDIER THE INSTANT HE IS DROPPED BY A BULLET THROUGH THE HEAD IN FRONT OF CORDOBA

DEATH IN SPAIN: THE CIVIL WAR HAS TAKEN 500,000 LIVES IN ONE YEAR

On July 17 the Spanish Civil War will be one year old. In that time it has brought Death to 500,000 Spaniards, has shattered such ancient cities as Madrid, Toledo, Bilbao, Irun and Durango, has kept Europe in a state of jitters.

When the war started, most U. S. citizens looked on the Loyalists as a half-crazy, irresponsible, murderous scum that had turned on its honorable betters. A year of war has taught the U. S. more of Spain.

The ruling classes of Spain were probably the world's worst bosses—irresponsible, arrogant, vain, ignorant, shiftless and incompetent. Some 20,000 landlords owned 50% of the land. They did not give their field hands modern machinery or their land modern irrigation. They refused to rent unused land to landless peasants for fear of giving the peasants dangerous ideas of ownership. The land was only about 25% efficient and much of it was idle. And Spain's mineral resources, among the greatest in Europe, lay almost entirely unexploited. The aristocracy of Spain was still living on the interest on wealth brought home from the Americas by the gold fleets in the 16th Century.

To the 20,000 landlords, add 21,000 Army officers, more than twice the total of British Army officers. There was one officer for every six privates, one general for every 150 men. For every $3 spent on soldiers' pay, food, barracks, ammunition, officers got $10 in pay—25% of the national budget. The national law made officers semisacred. Just for pushing a policeman of the swank Civil Guard, six Americans got six months in jail in 1935.

Add to the 41,000 landlords and officers, 100,000 clergy, the most top-heavy Church hierarchy in the world, next to Tibet. These also were paid by the State. The Church, with its enormous wealth, naturally took a capitalist's position. It was up to its neck in politics. Peasants were told that to vote against the Conservatives was usually a mortal sin. The Church was in charge of Spanish education. Result: the Spanish people were 45% illiterate. The reason for the civil war was simply that the people of Spain had fired their bosses for flagrant incompetence and the bosses had refused to be fired.

For a new movie of the Spanish war from the Government side, turn page.

star appeal, nor the continuing currency of his brand of war photography in photographic circles.[38]

Although the so-called *Falling Soldier* image has subsequently been reproduced in innumerable anthologies of war photography, it is instructive to view it in one of its original material contexts: *Life* magazine, 12 July 1937. Positioned next to a full-page advertisement for Vitalis hair tonic for men, Capa's image forms part of a consumer economy of images, presumably appealing to the kind of sporty male readers that might be as interested in a conflict that 'has kept Europe in a state of jitters' as they would 'five sets of tennis in the broiling sun'. The curiously similar attire of the Loyalist soldier and American tennis player, both captured in dynamic attitudes of exertion, makes for a striking commentary on masculinity and the consumption of images in interwar America.

Truth and Authenticity

One of the earliest expectations of war photography was that it would reveal the 'truth' of war for the first time, offering what the *Practical Mechanic's Journal* of 1854 hoped would be 'undeniably accurate representations of the realities of war'.[39] To be sure, cameras captured a profusion of details in a mechanical and apparently objective manner. But are 'accurate representations' necessarily truthful? How did Fenton's rearrangement of cannonballs affect the essential authenticity of the scene? Does Capa's *Falling Soldier* communicate a certain truth about war, whether or not it actually depicts the moment of death? Can the image of 'Pancho' Villa leading an orchestrated cavalry charge be considered an accurate depiction of conflict? Questions of truth, objectivity and authenticity – long debated in the broader field of photography – become particularly charged when they concern the depiction of war.

The Australian photographer Frank Hurley courted controversy with his depictions of the Western Front, where he was official photographer to the Australian Imperial Force from 1917 to 1918. The First World War, in contrast to the Mexican Revolution, was not a particularly 'photogenic' conflict. Trench warfare, where opposing forces were half-buried and

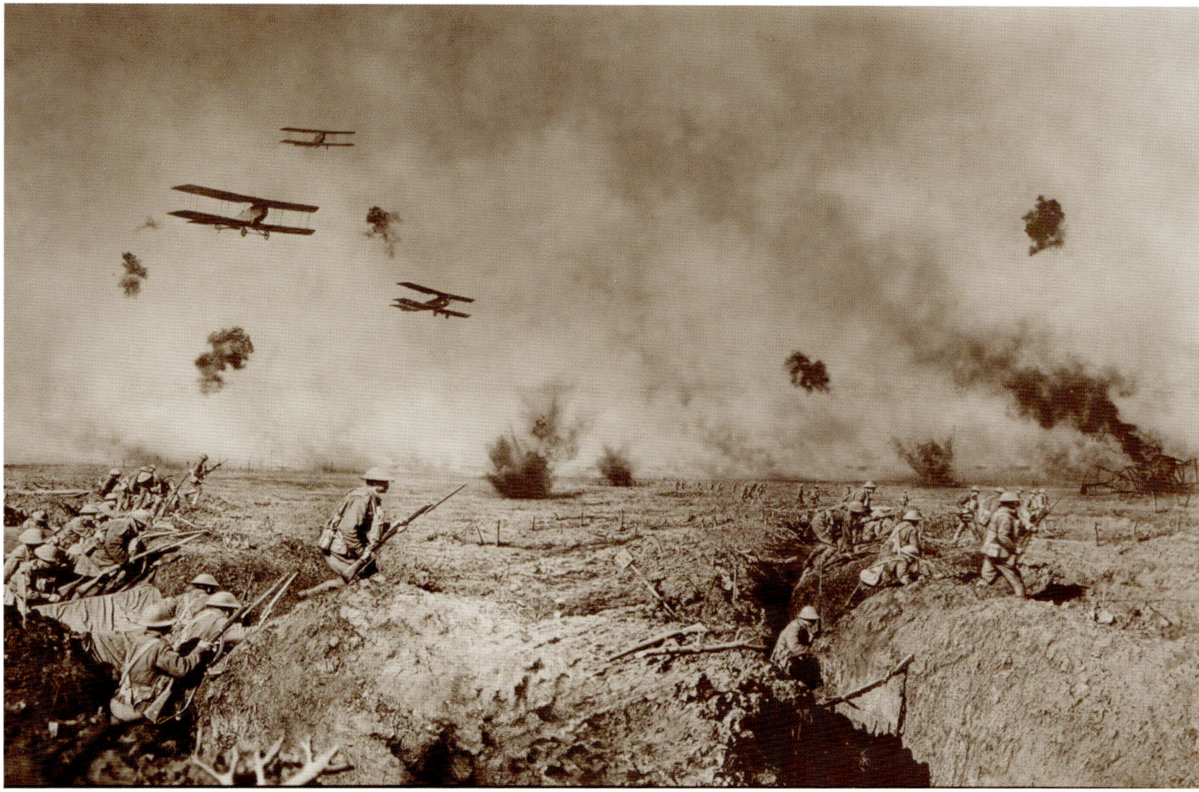

10 Capt. Frank Hurley, *An Episode after the Battle of Zonnebeke*, 1918, sepia-toned photograph on mount.

separated by no-man's-land, did not invite images of dashing cavalry charges. Sufficient illumination – a prerequisite for handheld cameras – was not in abundance in northern France and Flanders, and in any case much of the action took place at night. Hurley's diary records his photographic frustration with 'figures scattered, atmosphere dense with haze and smoke – shells that would simply not burst when required'.[40]

Nonetheless, Hurley produced an astonishing series of images representing the aerial and territorial battles of the First World War. Dissatisfied with the banalities of ordinary photographs, which he felt did not capture the true scale or experience of combat, Hurley produced composite images from multiple negatives. A typically bravura work, *An Episode after the Battle of Zonnebeke*, shows infantrymen with bayonets

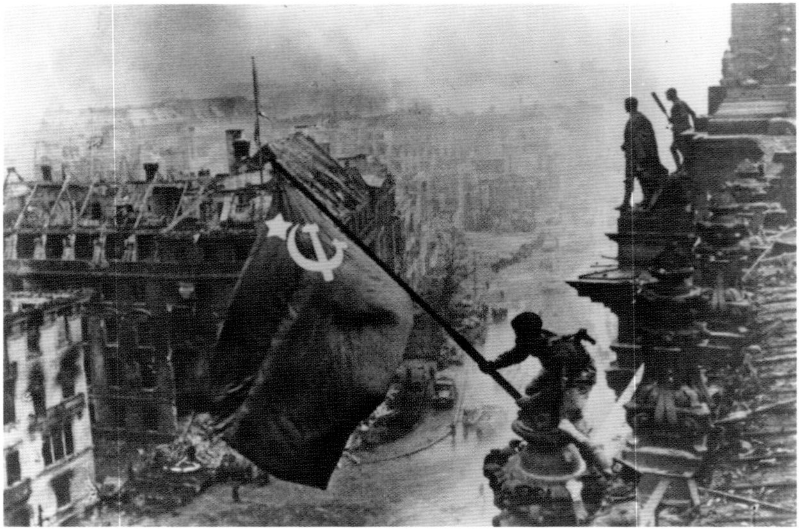

11 Yevgeny Khaldei, *Raising the Soviet flag at the Reichstag*, Berlin, 1945, silver gelatin print.

charging from the trenches against a hail of shell-bursts, while three fighter planes cross the smoke-streaked sky above (illus. 10). In the foreword to the catalogue for an exhibition of his work in 1918, Hurley declared,

> I make no claim to pictorial merit; the pictures are records, and except for several of the larger ones are faithful reproductions of the scenes they portray. In order to convey accurate battle impressions, I have made several composite pictures, utilising a number of negatives for the purpose.[41]

Hurley's conviction of fidelity was not shared by the Australian Imperial Force. He resigned following orders not to produce composites, and his London exhibition *Australian War Pictures and Photographs* attracted hostile reviews of his 'fake' pictures.[42] Despite his sincerity, Hurley's artistic composites were soon out of step with modernist photography and burgeoning notions of photojournalistic truth and objectivity of the 1920s.

Yet single-negative, single-moment images are not necessarily any more 'truthful' than Hurley's composites. In 1945 the Ukrainian photographer Yevgeny Khaldei made an image of a Russian soldier raising a flag atop the Reichstag in Berlin, after the city's surrender (illus. 11). The image is a heady combination of patriotism, victory and destruction, with the silhouetted figures and clouds of smoke making a powerful composition. The act of flag-raising has long held propaganda value, and Khaldei was commissioned by Soviet authorities in direct response to Joe Rosenthal's now-iconic *Flag Raising at Iwo Jima*, a morale-boosting image of victorious U.S. soldiers in the Pacific theatre, made a few months earlier. Khaldei, under the auspices of the Soviet news agency TASS, flew from Moscow to Berlin to make a comparable picture.[43] He reputedly came prepared with a Soviet flag made by his uncle from a red tablecloth and, working with a group of soldiers, orchestrated a re-enactment of a flag-raising that had occurred two days earlier. Further interventions were made in the preparation prints for publication. The contrast was enhanced in the darkroom, with the smoke and figures 'burned in' to achieve a more dramatic effect. As a concession to Soviet censors, Khaldei removed one of the two wristwatches worn by the man in the bottom right-hand corner to avoid any suggestion of looting, an ignoble (not to mention punishable) offence. The image was first published in the Russian magazine *Ogoniok* on 13 May 1945, and has since become a classic of war photography. The constructed nature of Khaldei's photograph, however, has been a topic of critique. Khaldei persistently maintained that his image represented the essential 'truth' of a historic event, albeit not one that photojournalist purists would recognize.

The Ethics of Images of Suffering

Photography's capacity to transmit graphic images of violence to spectators has been a persistently troubling issue. Some of the first photographs of battle fatalities were made by Corfu-born Felice Beato, who travelled extensively in India and the Far East, photographing the Indian Rebellion (1857) and the Second Opium War (1856–60). His image *Interior of Taku Fort* (1860) depicts the bodies of Chinese soldiers who have been killed by Anglo-French forces (illus. 12). An eyewitness to the scene recalled,

> [The ramparts] were thickly strewed with dead – in the north-west angle thirteen were lying in one group around a gun. Signor Beato was here in great excitement, characterising the group as 'beautiful', and begging that it might not be interfered with until perpetuated by his photographic apparatus, which was done a few minutes afterwards.[44]

The corpse-strewn image demonstrates Beato's talent for capturing powerful, even elegant, scenes of destruction. Like Roger Fenton, he was willing to intervene to produce more effective compositions, and was even reputed

12 Felice Beato, *Interior of Taku Fort*, 1860, albumen print.

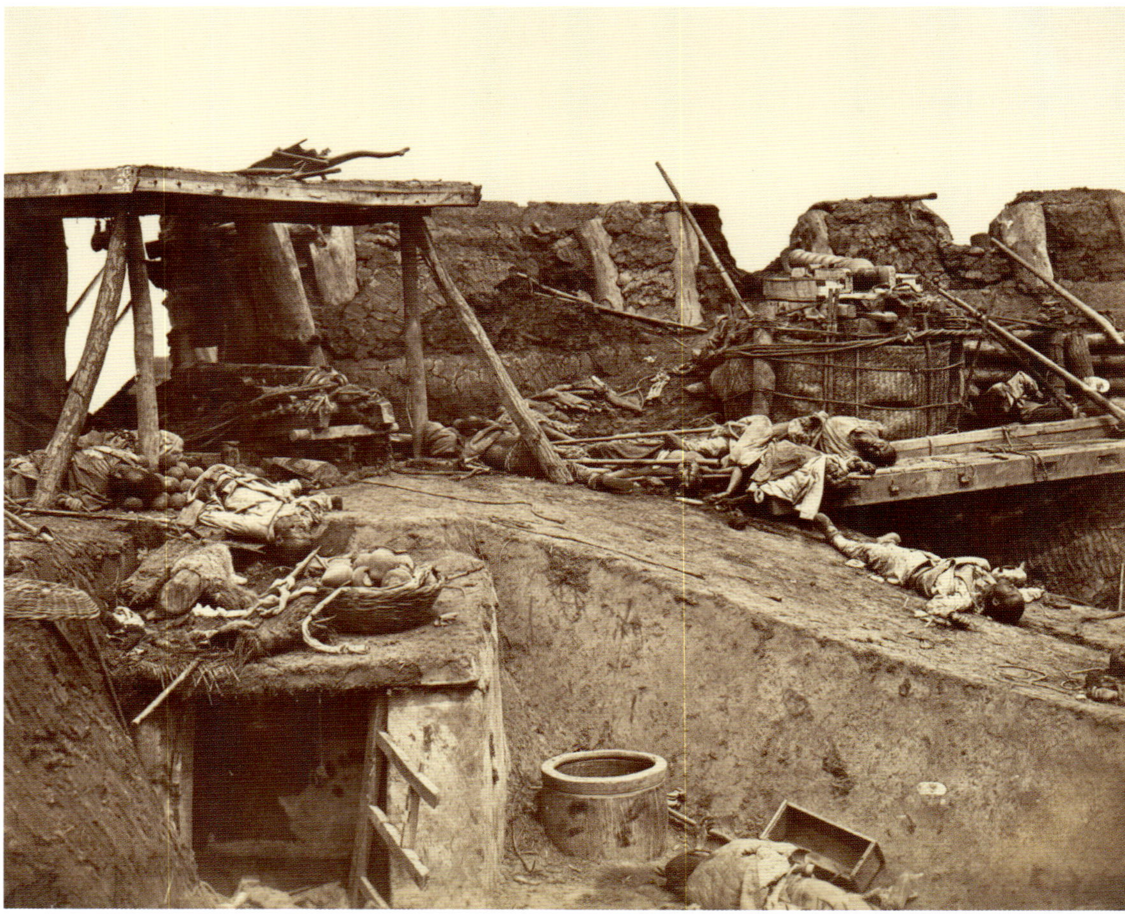

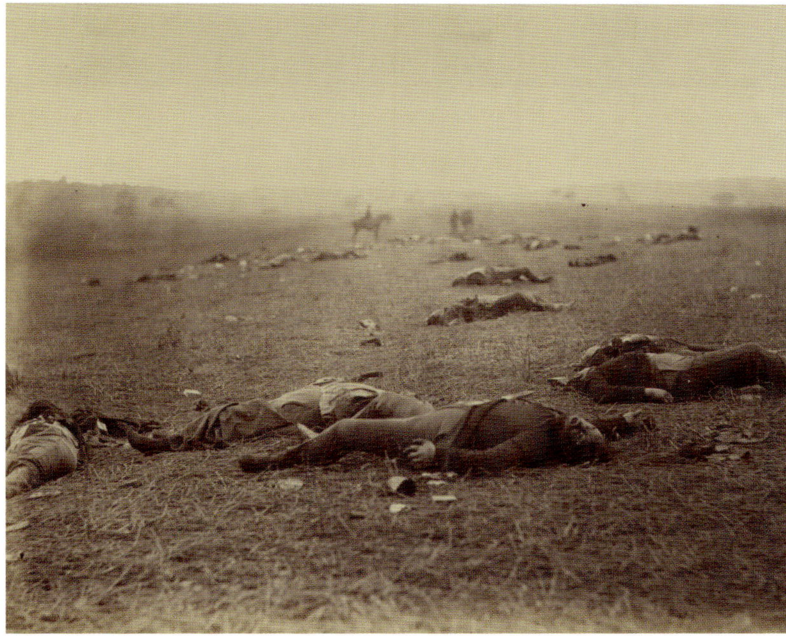

13 Timothy O'Sullivan, *A Harvest of Death, Gettysburg, Pennsylvania*, July 1863, albumen print in *Gardner's Photographic Sketch Book of the War* (1866).

to have rearranged corpses and disinterred bones for artistic effect.[45] Beato's detachment from the suffering of foreign bodies may be repugnant to present-day viewers, but it is consistent with the ideology of Western imperialism of the epoch.

Photographs of dead soldiers, often explicitly gruesome, were circulated widely during the American Civil War. Viewers were ambivalent in their responses, expressing both revulsion and attraction. The U.S. writer Oliver Wendell Holmes cautioned viewers of the 'strange spell' and 'terrible fascination' exerted by photographs of soldiers killed at Antietam, on display at Brady's Gallery in New York.[46] The British journalist George Augustus Sala described a horrific image of a soldier, disembowelled by a shell, whose entrails were 'most beautifully depicted'.[47]

The tensions between beauty and horror are palpable in Timothy O'Sullivan's *A Harvest of Death*, a photograph that graphically depicts rain-soaked casualties after the three-day Battle of Gettysburg in July 1863 (illus. 13). The image was disseminated in a number of formats, but none so resplendent as the two-volume album *Gardner's Photographic Sketch Book of the War* (1866).[48] Produced by Brady's erstwhile photographer and business partner Alexander Gardner, the weighty leather-bound tomes were produced with gilt-edged pages and marbled end papers, containing fine albumen prints made by Gardner himself. The term 'Sketch Book' used in the title, together with an elaborate frontispiece by *Harper's Weekly* illustrator Alfred Waud, presents Gardner's album as an artistic enterprise, and positions its contents within a longer history of romanticized imagery of battles in paintings and engravings.

The art-historical framing and the poetic language of the accompanying texts are, however, at odds with the photographs depicting battlefield corpses. *A Harvest of Death* presents a series of lifeless bodies, one of which exhibits a bloated and distorted face to the viewer. Here is conflict depicted not through the filter of painting or illustration, Gardner suggests, but through the mechanical gaze of the camera, delineating 'the dreadful details' in all their horror. The accompanying text declares that the image 'shows the blank horror and reality of war in opposition to its pageantry', a warning to future warmongers.

It has frequently been asserted that images of suffering might prevent or end conflict, most notably in relation to the Vietnam War (1955–75). The Cold War proxy conflict saw South Vietnam forces, supported by U.S. troops, struggle against Soviet-backed North Vietnam. The conflict was comprehensively covered by independent, free-ranging photographers, and many of war's most hard-hitting and enduring icons were made in the conflict, among them Larry Burrows's *Reaching Out* (1966), Philip Jones Griffith's *Civilian Victim* (1967), Eddie Adams's *Saigon Execution* (1968), and *Accidental Napalm* (1972) by Nick Ut, the professional name of Vietnamese photojournalist Huỳnh Công Út. This last image shows a group of children fleeing ominous clouds of smoke, their faces contorted in pain (illus. 14). At the centre is a screaming naked girl, her arms outstretched in a gesture of supplication, while the camouflage-clad soldiers behind her seem curiously unmoved. Like many icons of war photography, the image has accrued layers of (mis)interpretation, such as the assumption that U.S. aircraft were responsible for the erroneous napalm strike on South Vietnamese civilians (it was, in fact, carried out by South Vietnamese troops). The connotation

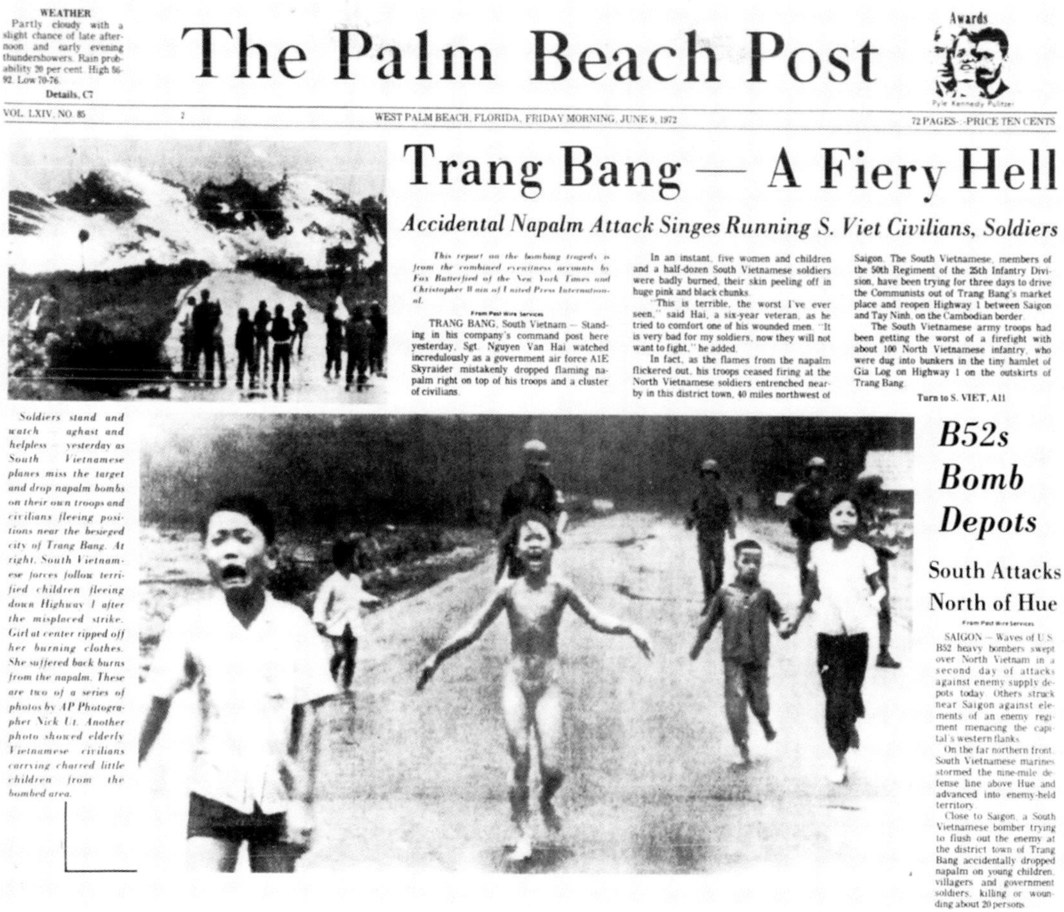

WEATHER
Partly cloudy with a slight chance of late afternoon and early evening thundershowers. Rain probability 20 per cent. High 86-92. Low 70-76.
Details, C7

The Palm Beach Post

Awards
Pyle Kennedy Pulitzer

VOL. LXIV, NO. 85 | 2 | WEST PALM BEACH, FLORIDA, FRIDAY MORNING, JUNE 9, 1972 | 72 PAGES--PRICE TEN CENTS

Trang Bang — A Fiery Hell

Accidental Napalm Attack Singes Running S. Viet Civilians, Soldiers

This report on the bombing tragedy is from the combined eyewitness accounts by Fox Butterfield of the New York Times and Christopher Wain of United Press International.

From Post Wire Services

TRANG BANG, South Vietnam — Standing in his company's command post here yesterday, Sgt. Nguyen Van Hai watched incredulously as a government air force A1E Skyraider mistakenly dropped flaming napalm right on top of his troops and a cluster of civilians.

In an instant, five women and children and a half-dozen South Vietnamese soldiers were badly burned, their skin peeling off in huge pink and black chunks.

"This is terrible, the worst I've ever seen," said Hai, a six-year veteran, as he tried to comfort one of his wounded men. "It is very bad for my soldiers, now they will not want to fight," he added.

In fact, as the flames from the napalm flickered out, his troops ceased firing at the North Vietnamese soldiers entrenched nearby in this district town, 40 miles northwest of Saigon. The South Vietnamese, members of the 50th Regiment of the 25th Infantry Division, have been trying for three days to drive the Communists out of Trang Bang's market place and reopen Highway 1 between Saigon and Tay Ninh, on the Cambodian border.

The South Vietnamese army troops had been getting the worst of a firefight with about 100 North Vietnamese infantry, who were dug into bunkers in the tiny hamlet of Gia Log on Highway 1 on the outskirts of Trang Bang.

Turn to S. VIET, A11

Soldiers stand and watch — aghast and helpless — yesterday as South Vietnamese planes miss the target and drop napalm bombs on their own troops and civilians fleeing positions near the besieged city of Trang Bang. At right, South Vietnamese forces follow terrified children fleeing down Highway 1 after the misplaced strike. Girl at center ripped off her burning clothes. She suffered back burns from the napalm. These are two of a series of photos by AP Photographer Nick Ut. Another photo showed elderly Vietnamese civilians carrying charred little children from the bombed area.

B52s Bomb Depots

South Attacks North of Hue

From Post Wire Services

SAIGON — Waves of U.S. B52 heavy bombers swept over North Vietnam in a second day of attacks against enemy supply depots today. Others struck near Saigon against elements of an enemy regiment menacing the capital's western flanks.

On the far northern front, South Vietnamese marines stormed the nine-mile defense line above Hue and advanced into enemy-held territory.

Close to Saigon, a South Vietnamese bomber trying to flush out the enemy at the district town of Trang Bang accidentally dropped napalm on young children, villagers and government soldiers, killing or wounding about 20 persons.

14 Nick Ut (Huỳnh Công Út), *Accidental Napalm*, reproduced on the front page of the *Palm Beach Post*, 9 June 1972.

of the image, however, was that U.S. military intervention in a far-off country had directly caused the suffering of innocent children. It has long been commonly held that Ut's image, together with those of his colleagues Burrows, Jones Griffiths, Adams et al., were instrumental in turning public opinion in the U.S., leading to popular and congressional opposition and, eventually, the end of the war. Subsequent scholars have been more circumspect: such definite cause-and-effect proves almost impossible to determine, and images may be better described as having reflected or crystallized the nation's mood as it was turning against the war.[49]

The Vietnam period arguably represents the acme of 'classic' war photography, but the participation of some photographers seems to have been configured as a macho rite of passage in an exotic country. The British photographer Tim Page has been especially honest in his descriptions of war as 'rock 'n' roll', his participation motivated by a predilection for drugs and military hardware rather than a loftier mission to depict the 'truth' of war. *Tim Page's Nam* (1983) presents the war photographer as a kind of rock star whose own experience, as the title and cover blurb suggests, is at least as important as those of the people depicted:

> Tim Page has become part of the Vietnam War mythology . . . he soon became identified with a particular 'cool', buccaneering attitude to the war . . .Wounded five times, the last time nearly fatal, he was himself as much a war victim as the military and civil personnel whose portraits appear in this book.[50]

Along with U.S. photojournalists Sean Flynn and Dana Stone, Page is thought to be one of the inspirations for the character portrayed by actor Dennis Hopper in the seminal Hollywood film *Apocalypse Now* (1979), which took the stereotypical persona of 'crazy photojournalist' to its logical conclusion. The war photographer had become a glamorous cultural figure, but was also morally ambivalent.

By the 1980s, conventional war photography clashed with new thinking from post-structuralist, postcolonial and feminist viewpoints. In 1981 the influential American photo-artist and theorist Martha Rosler critiqued the shortcomings of photo-documentary, deploring the tendency of professional photographers (explicitly naming Robert Capa) to 'parachute in' to far-off war zones. The focus on dramatic events and the war photographer's celebrity status, Rosler claimed, told the viewer nothing about the systemic relations between power, politics and war.[51] She favoured practices that disrupted the illusory realism of the documentary photograph, such as photomontage (see illus. 67). The problematics of exploitation, voyeurism and 'beautiful suffering' (to use the term coined by Reinhardt, Edwards and Duganne) gained wider currency in the late twentieth century and continue to fuel debate.[52] Images of war, viewed from a safe distance, may invite pleasurable aesthetic contemplation, whereas those photographed may be in positions of vulnerability, unaware that they are being photographed, or unable to give consent.

Steve McCurry's iconic image *Afghan Girl* (1984) has been a focus of these concerns. Depicting a young war refugee from Soviet-occupied Afghanistan, the photograph has become one of the most recognizable and popular images published by *National Geographic* magazine. The striking green eyes of its subject have stared out from the book jacket of McCurry's hardback monograph *Portraits*; the cover of *National Geographic: 100 Best Pictures: Special Member's Edition*; and, at the time of writing, a poster bearing McCurry's name in capitals, sold in support of the charity ImagineAsia

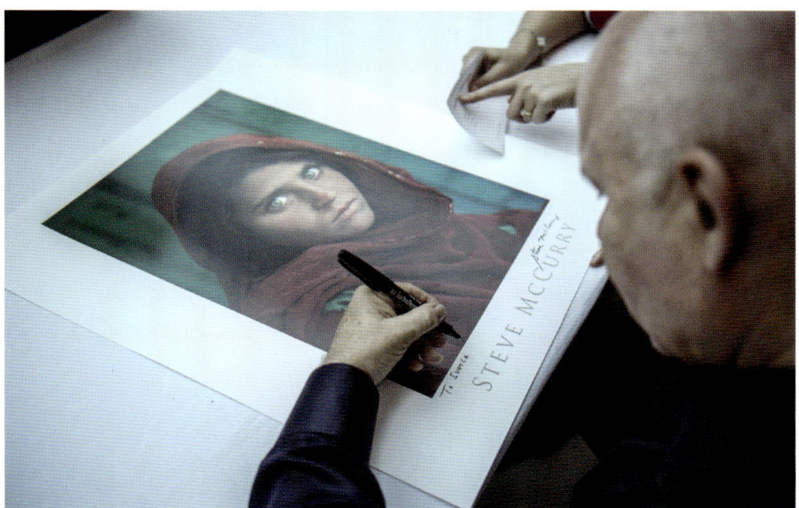

15 Mustafa Kamaci, Steve McCurry signing a poster of his 1984 photograph *Afghan Girl*, 2015.

which he co-founded in 2004 (illus. 15).[53] McCurry received international acclaim for his ability to depict 'humanity's commonality' and 'the true universality of the depths of human emotion', and was elevated to the pantheon of Magnum photographers.[54] On the other hand, as Holly Edwards has argued, the real-life subject of *Afghan Girl* remained overlooked, although her image helped raise a considerable amount of money for charity, as well as aiding *National Geographic* to sell books and magazine subscriptions.[55] When U.S. struggles with the Taliban regime put Afghanistan back on the public radar, *National Geographic* mounted a campaign – supported, not coincidentally, by the FBI – for Steve McCurry to locate his 'enigmatic' subject. Sharbat Gula was eventually 'found' in 2002, unaware of the use of her image, from which she had not personally profited. When interviewed, she remembered her anger at the intrusion of a Western male photographer into the girls' school tent where, for the first time in her life, she was photographed.[56] This is not to suggest callousness on the part of McCurry, who has gone to considerable lengths to support charitable work in Afghanistan, but Gula has clearly lacked control over her representation, both at the moment of its production and its subsequent dissemination. The case of *Afghan Girl* highlights some of the ethical dilemmas and contradictions of humanitarian photography, and the power imbalances that might result even from well-intended enterprises.

A further concern is that overexposure to images of suffering deadens the viewer's response, leading them to turn away rather than act. The notion of 'compassion fatigue' was first suggested by the U.S. cultural critic Susan Sontag in her seminal book *On Photography* (1977) and expanded by Susan D. Moeller in *Compassion Fatigue: How the Media Sell Disease, Famine, War and Death* (1998). Although an influential theory, it is unclear if things are that simple. Images of suffering – such as the drowned Syrian refugee Alan Kurdi (see illus. 65) – continue to be an important part of the media economy and charitable campaigns, suggesting that at least some viewers do more than merely glance and look away.[57] Photojournalists' images of war remain powerful, if unpredictable in their effects.

Alternative Viewpoints: Non-professionals

Professional war photographers, photojournalists and official military photographers are not, of course, the only ones to have made records of combat action and armed conflict. In 1888 George Eastman introduced the Kodak, proclaiming 'You push the button, we do the rest.' By the early twentieth century the mass availability of cheap, easy-to-use cameras, coupled with processing services, meant that unskilled observers could record their own viewpoints.[58] In the First World War camera manufacturers capitalized on this new market for snapshot photography. The German company Ernemann promoted its products to servicemen, claiming 'the best results in all theatres of war' (illus. 16). Eastman Kodak, meanwhile, re-branded its Vest Pocket model as 'The Soldier's Camera'.[59] With a breezy optimism that proved ill-founded, an advertisement from a New Zealand camera store in 1915 declared,

> The gift he will most appreciate is a Soldier's Kodak. It will enable him to bring back an invaluable picture-record of the Great War. Every incident in camp, aboard ship, in foreign ports, or at the front, makes an interesting subject for a Kodak Snapshot.[60]

The possibility of soldiers making their own photographs of their experiences was not welcomed by most of the belligerent authorities; Lord Kitchener, the

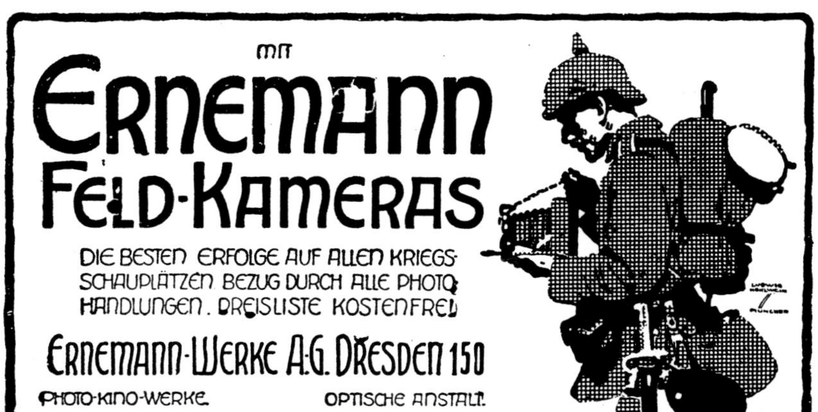

16 Advertisement for Ernemann Feld Kameras in *Woche*, 10 August 1918.

Secretary of State for the British government, declared that he would court-martial any man found in possession of a camera.[61] In practice, prohibition was impossible to enforce, and many thousands – possibly millions – of snapshots were made by British, French, Belgian, German, Russian and other soldiers for their own private use.

Photographic censorship, though by no means new, became institutionalized during the First World War when all belligerents rapidly established centralized censorship offices.[62] The motivations and methods for the control or suppression of images in wartime are complex and varied, determined by cultural, political and historical contexts. The control of military intelligence is of paramount consideration, but censorship goes much broader. Since Roger Fenton took his photographic apparatus to the Crimea in 1855, governments and the armed forces have considered it expedient to permit or to occlude certain images, and to control the effects they might have on soldiers, citizens and enemies. Censorship operates on varying levels: it may be legally sanctioned or illegally enforced; regulated by coercion, persuasion or indeed, a mixture of the two. Photographers and publishers may self-impose censorship, adhering to overtly stated policies by state or media agencies, or conforming to more tacit assumptions of taste, national loyalty or public appetite. Censorship, of course, goes hand-in-hand with propaganda, discussed in Chapter Three, 'Home Fronts'.

During the First World War, images that might demoralize soldiers or civilians at home were discouraged, if not always prohibited outright. Graphic images of the corpses of comrades rarely appeared in the press unless it was politically expedient for them to do so. A German photograph obtained by the British Sport and General Press Agency, for example, which shows the decomposing hand of a Belgian soldier protruding from a grave, was typical of images used to incite popular outrage among the British and U.S. publics.[63] More frequently, images of corpses remained private, kept by the soldiers and war workers who had made them.

Frustrated by censorship and restrictions on access to the frontlines, some periodicals advertised for snapshots made by soldiers, offering financial rewards. In a particularly notable instance, snapshots of the so-called 'Christmas Truce' of 1914 found their way into British newspapers, including the *Daily Mail*, *Daily Sketch* and *Daily Mirror* and the periodical

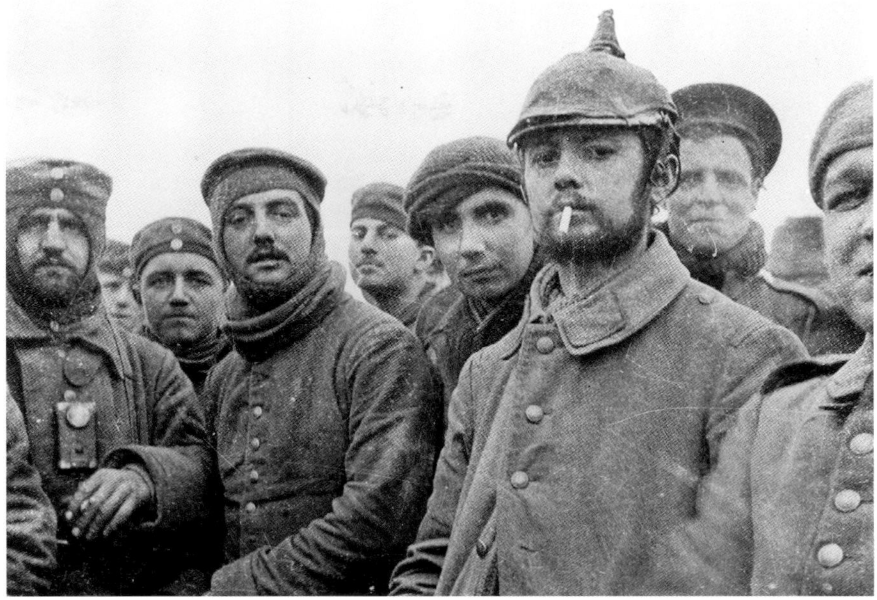

17 Private R. W. Turner (attrib.), British and German soldiers fraternizing at Ploegsteert, Belgium, on Christmas Day 1914, black-and-white snapshot.

The Graphic. The images, some of which have been attributed to Private R. W. Turner of the London Rifle Brigade, showed British soldiers fraternizing with troops of the 104th and 106th Saxon Regiments during an unofficial ceasefire at Ploegsteert, Belgium, on the Western Front (illus. 17).[64] Defying wartime agendas of national enmity, the images remain a powerful statement on individual humanity in the midst of the war machine.

Unofficial despatches from the combat zone were not restricted to male soldiers. Numerous women used snapshot cameras to record their experiences as nurses, volunteers, ambulance drivers and other auxiliary roles that placed them close to 'the action'. One such woman was Mairi Chisholm, a teenage Scot who, with her companion Elsie Knocker, helped establish and run a First Aid post near the trenches in Pervyse, Flanders.[65] Through photographs of shell bursts and battle casualties, Chisholm registered her proximity to danger and destruction. Other images, however, show moments of humour and reflect the vitality of people making the best of almost impossible circumstances. Knocker is pictured with Belgian soldiers playing on a rudimentary see-saw; another snapshot depicts Chisholm and friends rowing a boat in a rain-filled

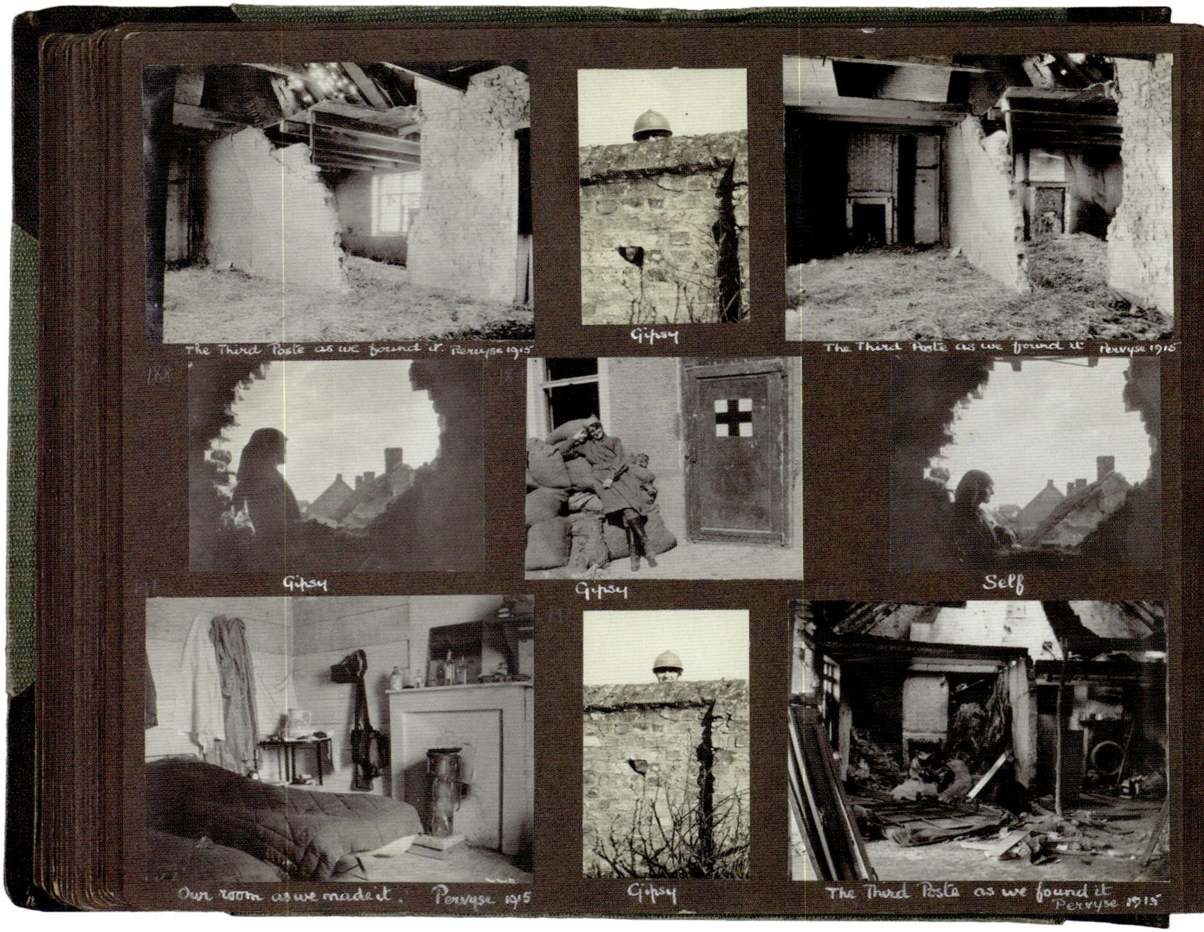

18 Mairi Chisholm, snapshots from Pervyse, northern Flanders, c. 1914–18, mounted in photo album with handwritten captions.

ditch, affectionately named 'The Punt at Henley'. Chisholm's photo albums are a jumble of such material, reflective of the contradictions of her precarious existence (illus. 18).[66] Such images offer a valuable supplement to the limited records that were sanctioned by state and military agencies.

War Photography as a Gendered Practice

Despite the many contributions of female photographers, war photography has been – and still largely is – conceived as a masculine, even macho practice. The persistent emphasis on close-combat action from the point of view of the combatant has proved a barrier to female participation. Since women have often been historically excluded from official combat positions (although in reality many have participated in fighting) they have found it harder to access front lines, or to be taken seriously as authentic reporters on war.

Most early photographers working in far-off combat zones were male. Women were handicapped by limited photographic education and employment, as well as restrictive perceptions of correct female behaviour. Nonetheless, some women found ways to document battle and its aftermath. The Indian-born British woman Harriet Tytler was in Delhi with her husband Major Robert Tytler when she was caught up in the 1857 Indian Rebellion against the British Empire.[67] Harriet was a keen artist, and it seems that she and her husband received photography instruction from none other than Felice Beato. Immediately after the conflict, the couple made around five hundred calotypes (an early form of silver-based photographic print) documenting former scenes of war, batteries and ruined buildings (illus. 19). Despite Harriet Tytler's fundamental role in producing these images, and her prowess as 'a most successful photographer', historians have frequently reduced her to the status of helpmate, sometimes omitting her name altogether.[68]

Numerous other women made early photographs in the context of war, such as Elizabeth Beachbard, Frances Benjamin Johnston and Olive Edis, all of whom feature in this book. By the 1930s, however, the genre of war photography was coalescing as a practice particularly suited to men and closely identified with the fighting soldier. The masculine war photographer

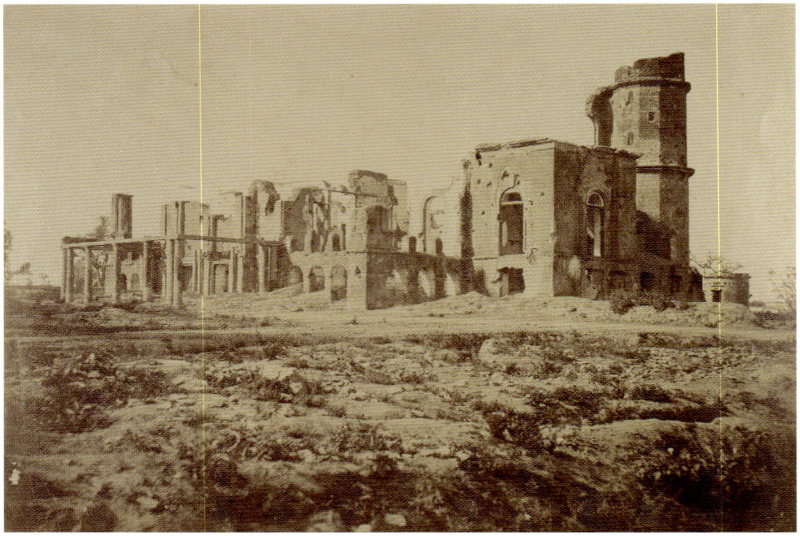

19 Harriet and Robert Tytler, *The Residency at Lucknow*, 1857, albumen print.

found its archetype, as photography scholar Patricia Vettel-Becker has argued, in the figure of Robert Capa.[69] Although he never served as a soldier, Capa's willingness to risk his life by going shoulder-to-shoulder with troops – most famously in the D-Day landings of 1944 – became an important part of his appeal. So too was his carefully cultivated, virile, dashing persona. He dated Hollywood star Ingrid Bergman, and claimed that '[for] a war correspondent to miss an invasion is like refusing a date with Lana Turner after completing a five-year stretch in Sing Sing.'[70] Yet 'Robert Capa' was originally a fiction: a marketing ploy by Gerda Taro and Endre Friedmann, who subsequently adopted the name permanently.[71] Taro made important contributions to the photography of the Spanish Civil War, with images featuring on the covers of *Vu*, *Regards* and *Die Volks-Illustrierte* (illus. 20). Known as *la pequeña rubia* ('the little blonde'), Taro found that her work was received differently from that of Capa, and after her untimely death in a tank collision in 1937 she was reduced to a footnote in Capa's romantic life. Thanks to recent investigation by Irme Schaber, Richard Whelan and others, Taro has been reassessed as a significant photographer in her own right.[72]

With the advances of second- and third-wave feminism, some concerted efforts have been made to alter the gender skew of war photography and what

20 Gerda Taro, 'In the Trench of Freedom', cover of *Die Volks-Illustrierte*, 25 August 1937, photogravure.

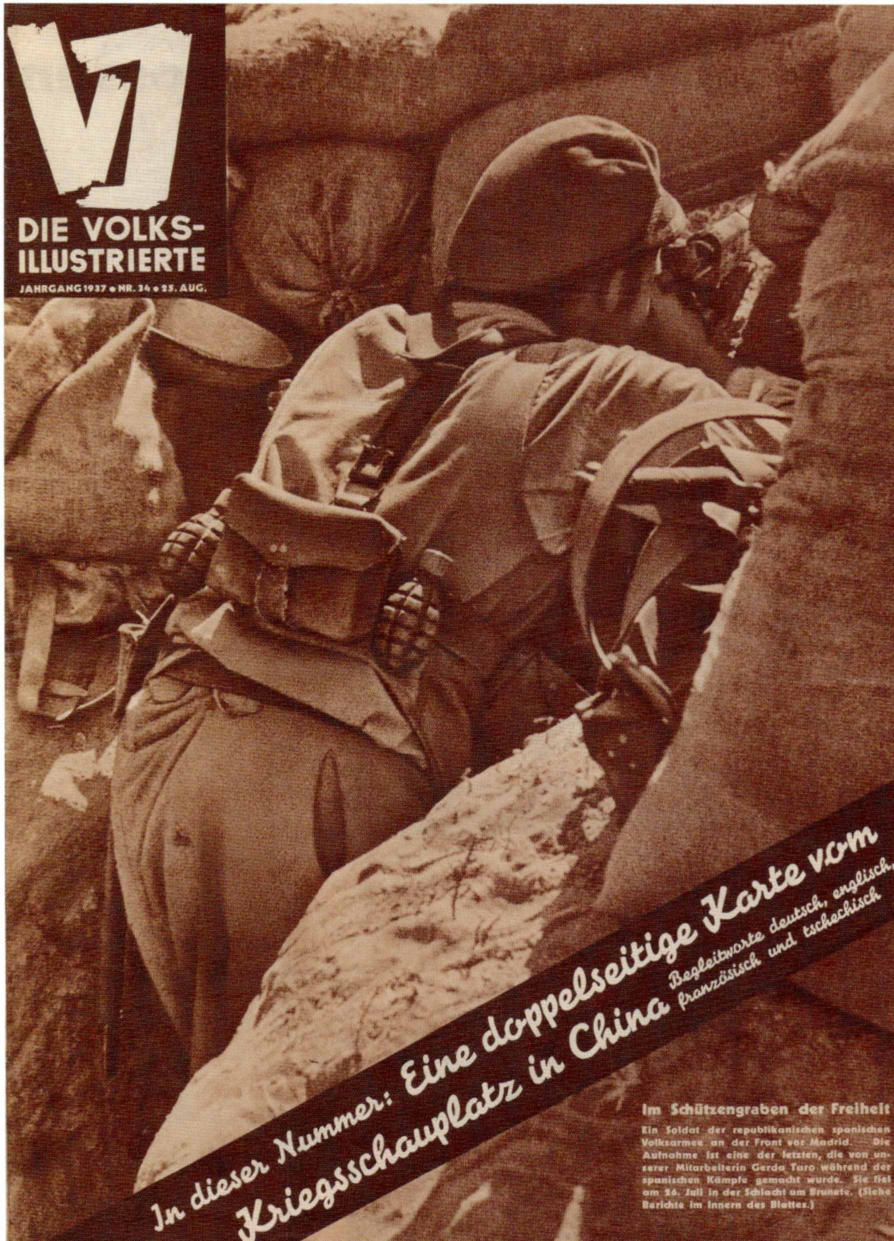
VI

DIE VOLKS-
ILLUSTRIERTE

JAHRGANG 1937 • NR. 34 • 25. AUG.

In dieser Nummer: Eine doppelseitige Karte vom Kriegsschauplatz in China

Begleitworte deutsch, englisch, französisch und tschechisch

Im Schützengraben der Freiheit

Ein Soldat der republikanischen spanischen Volksarmee an der Front vor Madrid. — Die Aufnahme ist eine der letzten, die von unserer Mitarbeiterin Gerda Taro während der spanischen Kämpfe gemacht wurde. Sie fiel am 26. Juli in der Schlacht um Brunete. (Siehe Berichte im Innern des Blattes.)

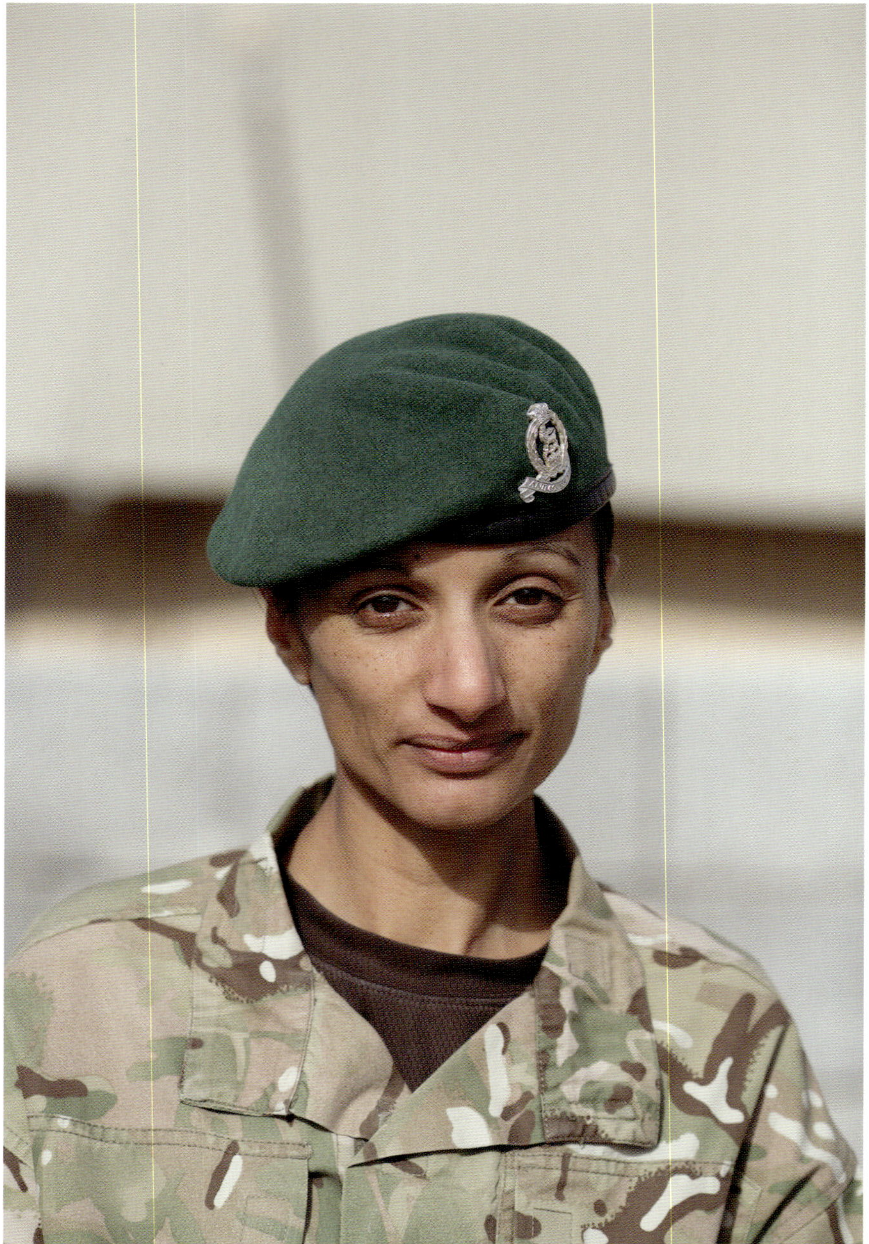

veteran war photographer Susan Meiselas remembers as the 'men's club' atmosphere of agencies like Magnum Photos.[73] In 1975 the U.S. not-for-profit group Professional Women Photographers was formed; the women-only agency Format was established in the UK in 1983. Much more recently, the world's first Middle Eastern all-female collective, Rawiya, was formed in 2009. Scholars and historians have sought to reassess the contributions of women such as Sara Castrejón, photographer of soldiers and firing squads in the Mexican Revolution; Käthe Buchler, an amateur on the German home front; and Lee Miller, the Second World War U.S. war correspondent for *Vogue*, all of whom are discussed later in this book.

The British photographer Alison Baskerville directly addresses questions of gender, the military and photographic history. Baskerville served for twelve years with the RAF in conflict zones including Iraq, and later worked as a military photographer for the British Army in Afghanistan. Subsequently re-training as a photojournalist, she produced *The Full Picture*, depicting women's lives in the British Army with an insider's insight. Baskerville was initially ignorant of the rich history of women photographing war. Her discovery of the work of Olive Edis, who photographed France and Flanders in 1919 at the request of the Imperial War Museum, was a revelation to her. Baskerville's *Soldier* (2011–16) pays homage to Olive Edis's skills in war photography, portraiture and photographic technology, particularly the colour autochrome process, which Edis helped pioneer.[74] In contrast to the dramatic images of combat action associated with war photography, Baskerville chose to make a series of portraits of women, mainly stationed in Afghanistan, whose roles range from logistics to frontline combat medics to artillery gunners (illus. 21). Presented as 'digital autochromes' in large LED lightboxes, the portraits have a commanding presence, but stop short of being unproblematically celebratory. These women are presented not as archetypal heroic warriors but as individuals whose expressions are ambivalent. *Soldier* may recognize women's participation in the armed forces, but Baskerville's portraits do not glorify war.

21 Alison Baskerville, *Bev*, from the series *Soldier*, 2011–16, colour transparency on LED light box.

Non-Western Viewpoints

It is important to recognize the active participation of photographers from beyond the U.S. and Western Europe. Fenton, Beato and the Tytlers were white Europeans who photographed colonial conflicts from the perspective of imperial outsiders. Yet indigenous people also capitalized on photographic technologies, and ongoing scholarship is helping to recover projects that have largely been excluded from Western histories of photography.

In South America, Uruguayan photographers documented the War of the Triple Alliance (1865–70), in which the combined forces of Uruguay, Argentina and Brazil routed those of Paraguay. Esteban García and Javier López were contracted by the photographic company Bate & Co. in the Uruguayan capital of Montevideo.[75] García photographed the aftermath of the siege of Paysandú in early 1865; the following year López went to the front line at the

22 Javier López, *Field Tent on the Battlefield*, 1866, albumen print.

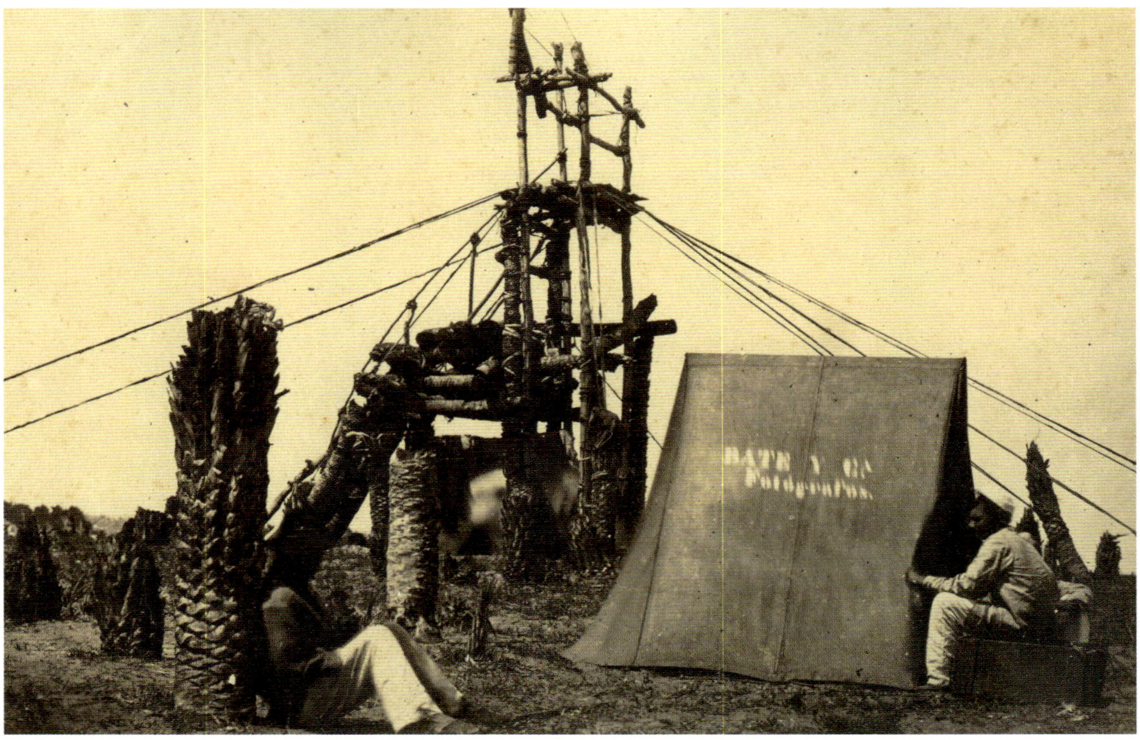

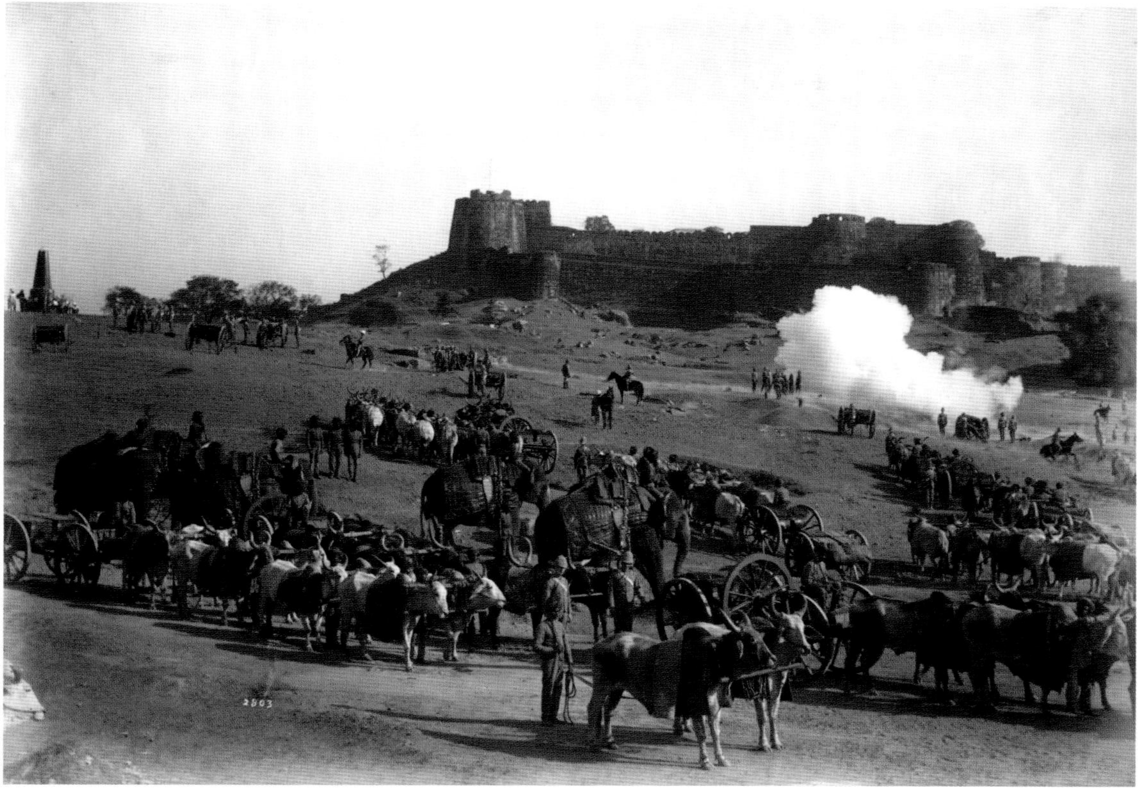

23 Lala Deen Dayal, Jhansi Fort and elephant battalions, Uttar Pradesh, northern India, c. 1885–7, albumen print.

Paraguayan-Argentine border. The swampy terrain meant transportation by cart was impossible, and López and his assistants carried photographic equipment on their backs. Using a field tent as a darkroom, they undertook the Herculean task of preparing, exposing and developing wet collodion plates (illus. 22), returning a second time to depict the devastating allied defeat at Curupayty. Despite being a commercial failure, the photographs of García and López bequeathed to posterity a unique record of a historic conflict.

The highly successful Indian photographer Lala (later Raja) Deen Dayal, portraitist to the maharajahs, made a number of records of military manoeuvres. One such image, from an album of albumen prints dated circa 1885–7, offers a majestic panorama of an elephant battalion in Uttar Pradesh, northern India, with Jhansi Fort as its apex (illus. 23). Dayal's image is beautifully composed, the viewer's eye led along the snaking line of troops and beasts of burden towards a carefully staged explosion, and finally the fortress itself. A native Indian who had originally trained as an engineer, Dayal displayed a talent for negotiating competing regimes of power, receiving patronage from both Indian elites and the British Raj.[76] His image reflects this ambivalence, celebrating the authority of the Imperial armed forces while also paying tacit homage to Rani Lakshmibai, defender of Jhansi Fort against the British in 1858, who became a heroine of national resistance.

The wartime participation of colonized or diasporic groups has, on the whole, been poorly represented. With a handful of exceptions, African Americans were rarely the central subjects of images made in the American Civil War (see illus. 29). The Indian Wars of the nineteenth century were documented from a U.S. perspective that configured Native Americans either as exotic and primitive warriors, or as dehumanized heaps of corpses inspected by dispassionate onlookers. There are precious few photographs of the North African soldiers among the Liberation of Paris in 1944, despite their decisive role in Charles de Gaulle's Free French Army.[77] In the face of archival lack, personal collections may offer alternative viewpoints. Lawrence McVey, who served in the 'Harlem Hellfighters' (369th U.S. Infantry), was one of the first African American soldiers to arrive in France, on 1 January 1918, and later received the Croix de Guerre medal for his participation in the conflict. Owing to institutional and individual racism in the U.S. Army, the regiment was segregated from others, and was assigned instead to the French Army.[78] A modest photographic postcard shows McVey standing to attention in military attire (illus. 24). The word 'HERO' has been added by an unknown writer, perhaps in defiance of the prejudice and discrimination experienced by McVey and his comrades, the photograph serving as a proud affirmation of service.

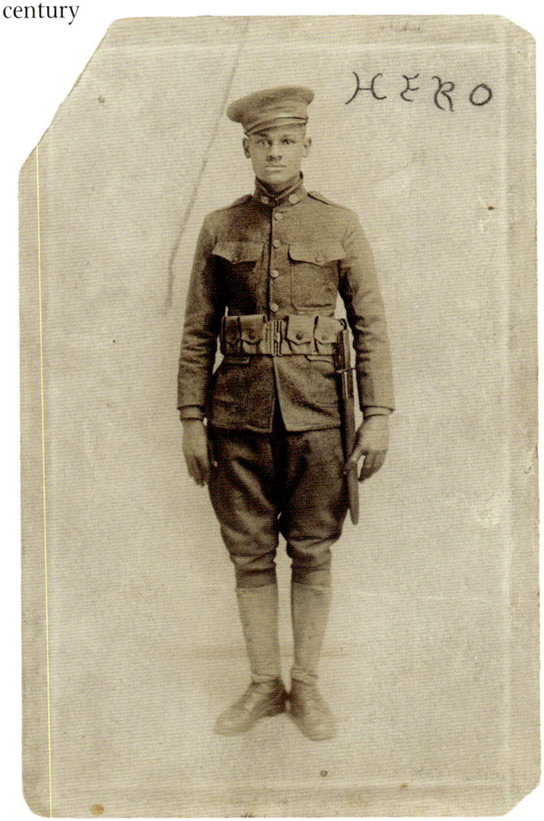

24 Alfred Philip Mighell, Lawrence McVey of the Harlem Hellfighters, c. 1914–18, annotated photographic postcard.

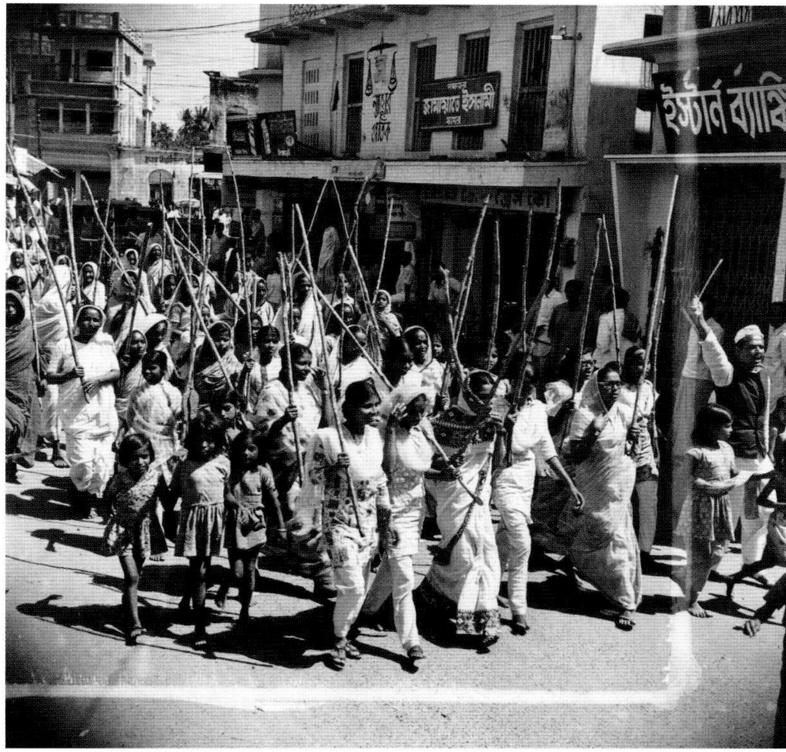

25 Mohammad Shafi, 'A rally with sticks or 'Lathi Michil', against the government of Yahya Khan. Garikhana Road, Jessore, Bangladesh', 23 March 1971, silver gelatin print.

In the late twentieth century, the mainstream media's reliance on Western photographers to supply news imagery from other parts of the world was increasingly challenged. Edward Said's 1978 landmark book *Orientalism* spearheaded a new postcolonial critique of Western perceptions of Asia, Africa and the Middle East, and the impulse to both exoticize and dominate other cultures. Mainstream media channels have been slow to recognize non-Western photographers, preferring the familiarity and convenience of their usual suppliers. In recent years, however, non-Western photographers have established their own agencies and collectives, aiming to supply international media with news images made by people who speak the language and are embedded in the socio-cultural context they depict.

The Bangladeshi agency and picture library Drik, established in 1989 by Shahidul Alam and Rahnuma Ahmed, not only represents indigenous

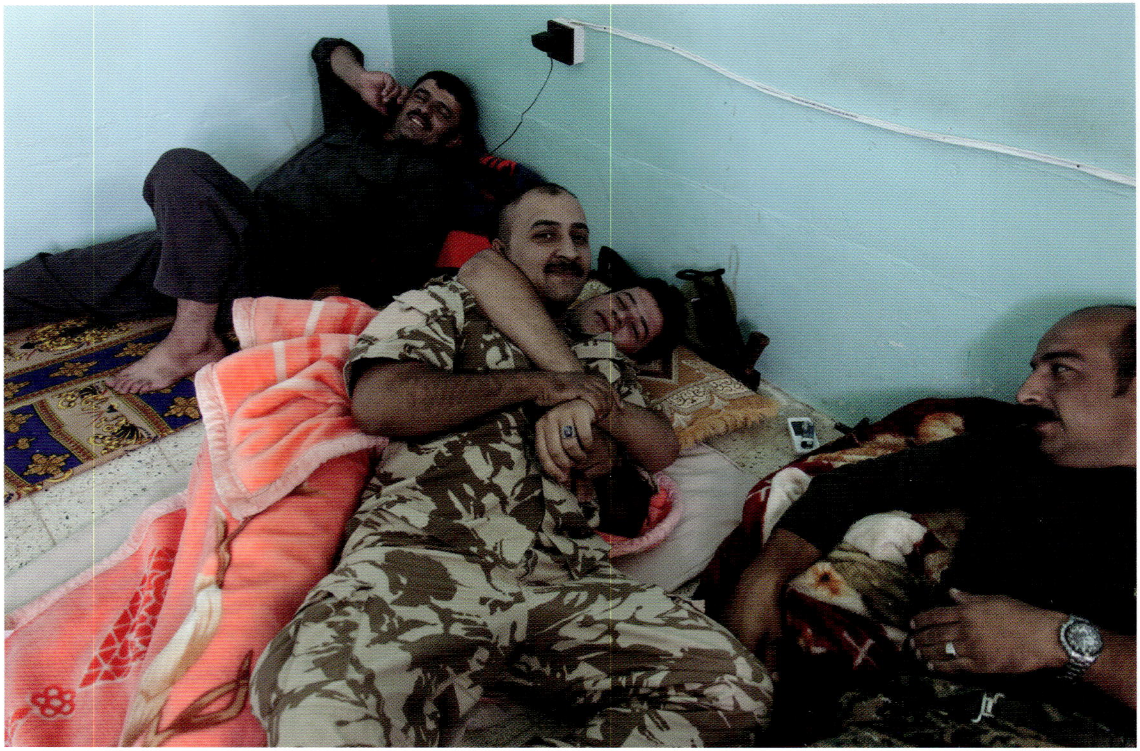

26 Bandan Atta, 'Peshmerga joke with each other at the base in Daquq', from the series *Diary of a Peshmerga*, 2014, colour digital photograph.

photographers but has retrospectively amassed an important historical archive of images made during the Bangladesh War of Independence (1971).[79] These images, made in defiance of the West Pakistani military regime, present insiders' views on a conflict that is inextricably linked to the partition of India in 1947 and Britain's colonial past. Subsequent exposure of these images, from international exhibitions to local tours via rickshaw, foregrounds the agency of Bangladeshi people in making their own history (illus. 25), in contrast to conventional Western images portraying helpless victims in need of what Alam calls a Western 'helping hand'.[80]

Recent developments in communications and digital technologies have hugely increased the viability of geographically dispersed agencies. In 2004 Drik founded Majority World, a web-based global platform for photographers from Asia, Africa, Latin America and the Middle East.

Other notable projects include Iraqi agency Metrography (2009), 3rd Eye in Afghanistan (2007) and Native, representing photographers from Africa and Latin America (2016). These agencies, and others like them, may collectively offer fresh viewpoints.

Bandan Atta, represented by Metrography, produced *Diary of a Peshmerga* in 2014. Most foreign photojournalists in Iraq have been 'embedded' – officially approved and attached to U.S. army units – and their images often sympathetically reflect the viewpoint of the U.S. soldier. By contrast, local photographer Atta documented the lives of *peshmerga* Kurdish Iraqi fighters at a military base in Daquq, Iraq, only a few miles away from Islamic State controlled territories. Atta clearly enjoys a rapport with the soldiers he photographs. In addition to more conventional images of combat action – including a spectacular shot of a rocket-propelled grenade being fired – his images include personal, even intimate, moments of camaraderie and vulnerability. Soldiers indulge in horseplay (illus. 26), perform mock drills with brooms and grin at the camera. Others invite Atta into their homes, permitting him to photograph them asleep, or with their children. Atta's work offers a more rounded record of the all-too-often dehumanized Western view of the Iraqi conflict. While agencies like Metrography remain only a minor part of the contemporary media economy dominated by corporations such as Getty Images and Alamy, they are helping change the ways in which wars are seen.

Although many of the individual photographers discussed in this chapter have been influential, they produce only a fraction of the images that are made in the context of war. More prolific by far are military, state and corporate entities that drive the mass production of photographic images and technologies, and which form an integral part of the war system. The nature of this 'military vision' is explored next.

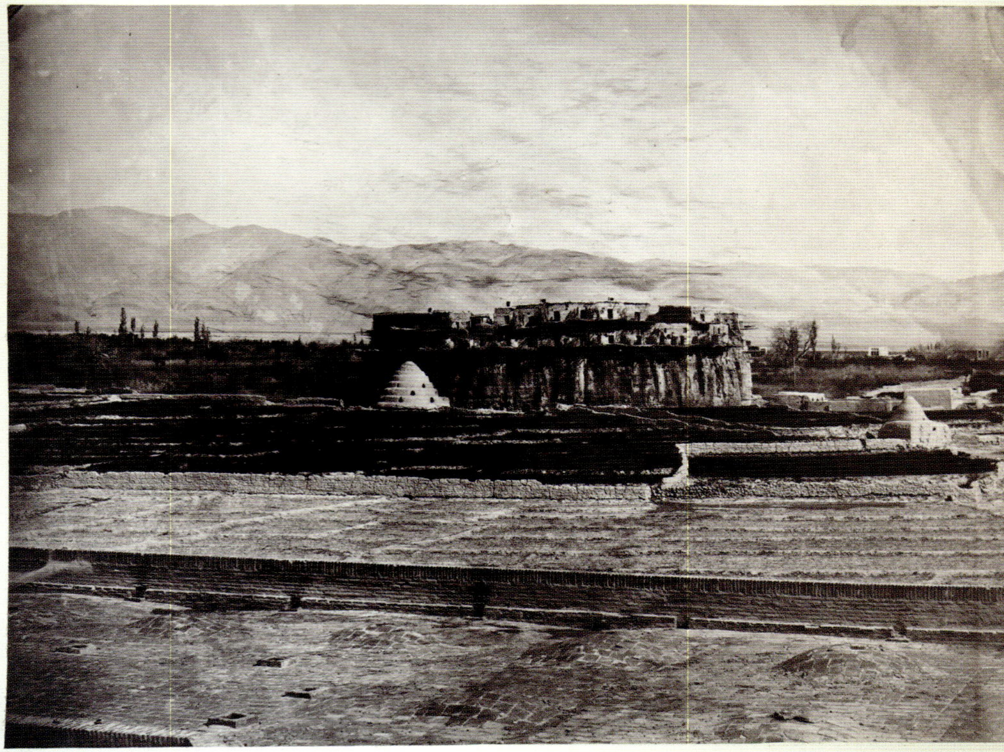

عکاسخانهٔ مدرسهٔ مبارکه دارالفنون دورنمای تپهٔ سید مصنوعی واقعه در لاسجرد که روی آن یکطبقه و دو طبقه بنا است و سابق تماماً مسکون بوده اینک **دارالخلافه ناصری کار عبدالله قاجار**

تقریباً پانزده خانوار سکنی دارند ارتفاع آن از سطح زمین تا انتهای عمارات دوازده ذرع و دورهٔ تل دویست و پنجاه قدم است اطراف تل آبادیهای جدید نموده اند باغات بسیار دارد آنچه خانه‌زاد بزحمت سیر نموده این است که از چاه آبی داخل قلعه رشته نقبی را سیاه کرد بعسرت با چراغ لانترن داخل شده صد قدم بواسطه صیقل مجری با سینه و دست رفته بعد فضا وسیع و خانها و انبارها و طویله‌ها دیده شد که دور تیز و محاذی خانهای روی تل و بعضی را انتهای نقب و مجرای آن بسته و مسدود است از قرار تقریر اهالی از این نقب بسیه یکفرسنگی این تل راه داشته که در ایام عسرت و تحصن آذوقه از نقب بالا میرسیده و راه معمولی پیچ در پیچ داشته که بسهولت اسباب و دواب را بنقب داخل نموده بعد راه را مسدود مینمودند اشخاصی که برای سد راه معمول در فوق بوده انداز این راه چاه آبی که خانه‌زاد داخل شده دو ذرع دور و هفت ذرع عمق دارد وارد شده خود را مصون و ایمن میداشتند یکصد و پنجاه خانوار جمعیت دارد

two

Military Vision

> By military necessity, the Army Air Forces have become the world's largest users of photographic film, paper, and equipment . . . More than 20,000,000 photographs a month are made for military purposes . . . In addition to reconnaissance photographs, other types of pictures are taken including training films and still pictures, still photographs of equipment, artillery fire control pictures, fighter gun camera pictures that reveal new types of air-craft in use by the enemy and the vulnerable points on these planes, and still other types of pictures for use in preparing maps and charts.
>
> '20 Million Photos a Month', *Science News-letter* (1945)[1]

The kinds of photographs described in this report do not conform to war photography in the popular imagination. None of these images was made by a professional photojournalist, nor has any won a prestigious award. There is no 'human interest' story to be communicated in these images, at least not in the traditional sense; nor do they convey any heroic drama on the battlefield. Yet this inventory of images is far more representative of the millions of photographs made for the purposes of war. Photographic technology – from aerial reconnaissance to infrared vision, X-rays to microfilm – has been, and remains, an essential part of an armed force's weaponry. This chapter explores some of the key ways in which photography, deployed as 'military vision', facilitates and enables warfare.

27 Abdollah Mirza Qajar, *Panorama of Lasjard*, c. 1883–96, annotated albumen print mounted in album.

Mapping and Surveying

In 1856 Eugène Durieu, co-founder of the Société Française de Photographie, presciently declared that photography would master unexplored countries in the same way that the victorious French Army had conquered lands.[2] Indeed, from the mid-nineteenth century onwards, photography was repeatedly used to survey territories and peoples in both a speculative and specific manner. Photography was one of a number of mid-nineteenth-century technological innovations (along with telegraphy, railways and steam navigation) that characterized the Industrial Revolution and an era of modernity, capitalism, exploration and colonialism.[3] Imperial expansion abounded, with projects led by the British in India and the Middle East; the French and Belgians in Africa; and the U.S. in the American West, among others. The camera's capacity for rapid documentation and surveying became an integral part of colonial expansion and control. The photographic survey was thus a hybrid of geographical, military and colonial impulses, part of a wider project to map, know and dominate terrains and peoples.

The French and British were among the first to incorporate photographic technology into the armed forces. The celebrated French portraitist Camille Silvy was an early advocate of the military application of photography and in 1859 was appointed war photographer for Napoleon III's campaign during the Second Italian War of Independence.[4] Silvy subsequently invented an early panoramic camera, which used a rolled waxed-paper negative, to aid in surveying and battlefield recording, perfecting his technique in 1867. Meanwhile, his compatriot Colonel Aimé Laussedat had been experimenting with photogrammetry (the science of measurement using photographic aids) and topographical photography. By 1861 the French War Ministry had formally adopted Laussedat's methods, and the civilian photographic innovator André-Adolphe-Eugène Disdéri – famed for his invention of the *carte de visite* portrait format – was employed to train officers in photography.[5]

Across the Channel, the British Royal Engineers, the military unit that had traditionally undertaken mapping, introduced photography instruction to their School of Military Engineering at Chatham in 1856. The School recognized the potential of photography to supplement traditional surveying techniques and record military engineering projects.[6] As part of the so-called

'Great Game', the geopolitical stand-off between the empires of Russia and Great Britain from 1830 to 1895, photography became increasingly fundamental to surveying contested territories in Central Asia. Civilian photographers such as John Burke were prolific suppliers of images, but the military also developed its own units. The Photograph School of the Bengal Sappers and Miners, a regiment of the Royal Engineers, made a number of photographs in northern Afghanistan during the Second Anglo-Afghan War (1878–80).[7] A series of albums entitled *Cabul* [Kabul] *Defences* depict fortifications constructed in 1880. Supplemented by text and plans, these photographic albums served as both confidential military gazetteers and records of work undertaken. A panoramic view, composed of two albumen prints joined together, depicts the *Bala Boorj* (upper tower) looking across the plains of Kabul (illus. 28). Made from an elevated position, the image communicates a sense of power and domination consistent with the later theories of French philosopher Michel Foucault, whose influential notion of an all-seeing central 'eye of power' asserts that territory may be controlled not only with weapons or armies, but through a gaze: 'a field of total visibility'.[8] Although the *Cabul Defences* albums have long been superseded as reliable military documents, photography's role as virtual 'watchtower' remains potent, via ever more sophisticated technologies.

28 Photograph School of Bengal Sappers and Miners, *Bala Boorj*, from the album *Cabul Defences*, c. 1879–80, albumen paper prints with gold toning, mounted in album with printed captions.

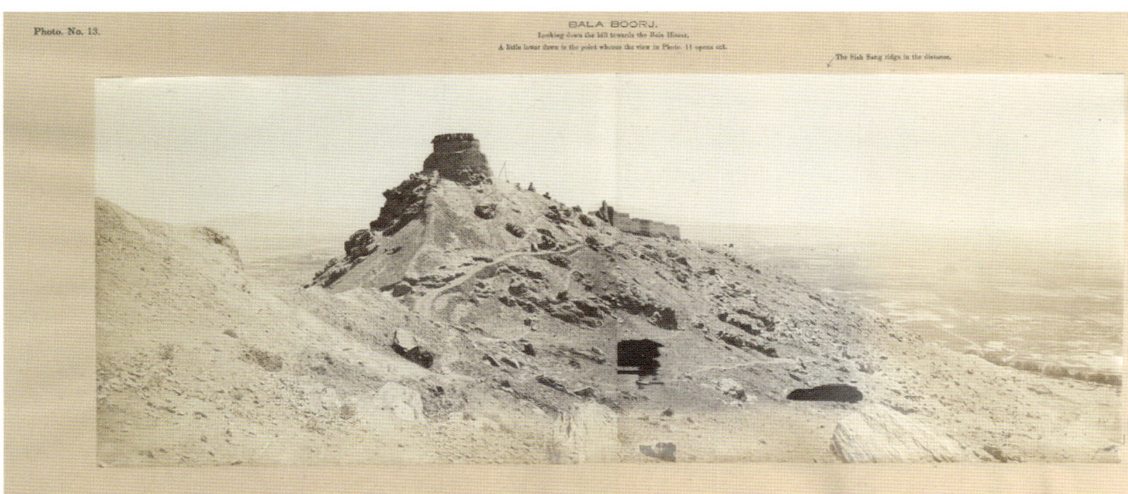

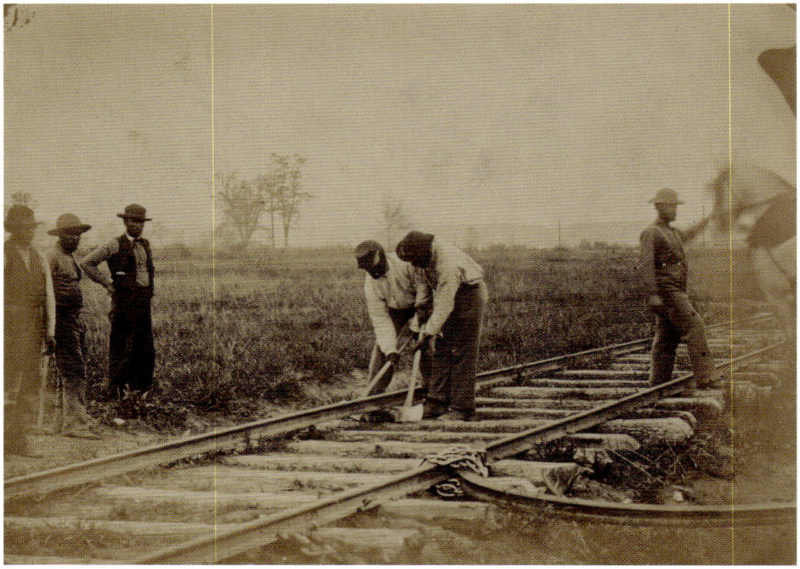

29 Andrew J. Russell, African American labourers work on a railroad in northern Virginia as part of the Union Army's military operations, *c.* 1862–3, salt print.

The U.S. military was initially slow to incorporate photography into its ranks, but the American Civil War promoted a boom in photographic surveys, essential to map terrain and record military installations. The Union had a distinct advantage over its Southern opponents, in that almost all photographic supplies were distributed in the U.S. through the New York-based company E. & H. T. Anthony & Co. A Federal blockade meant that the South resorted to smuggling chemicals under the guise of quinine, a permitted medicine, but the shortage severely limited Confederate photography.[9]

Surveys and documentation for the Union were often made ad hoc by civilian freelancers, some of whom worked for Mathew Brady's picture agency, most notably Alexander Gardner and Timothy O'Sullivan.[10] Official military photographers included Andrew J. Russell, who was assigned in 1863 to Brigadier General Herman Haupt, chief of the United States Military Railroads.[11] Russell's images were published for commanding officers in *Photographs Illustrative of Operations in Construction and Transportation* (1863), an instruction manual without pretensions to art or history. Beyond their immediate usefulness to the prosecution of war, Russell's records of military construction have preserved a portrait of a nation at a time of modernization.

His photographs of the construction of the Union Pacific Railroad, an essential supply line for the Union forces, are doubly important in their rare and prominent depiction of African American participation in the war (illus. 29).[12]

Photographic cartography became an invaluable tool to both sides, which were hampered by the scarcity of maps and poor knowledge of the terrain in which they were fighting. At the outset of war, General Richard Taylor complained that 'the Confederate commanders knew no more about the topography of the country than they did about Central Africa.' Survey parties were despatched to collect data, but reproduction of maps in the field proved problematic. It was impractical to transport the unwieldy stones and presses required for lithography, and tracing maps by hand was far too slow. Captain Albert H. Campbell, head of General Lee's Topographical Department, conceived a solution: tracing maps in Indian ink onto ordinary tracing paper, which were then used on photographic printing-out paper in sunlight. The resulting 'photocopies' were sometimes pasted together, or mounted on cloth, then folded to fit into pockets or saddlebags.[13]

The technique was also adopted by the Union.[14] A creased and dog-eared map, made on 13 May 1864, shows the topography of Resaca and encampments near Snake Creek Gap, Georgia, shortly after securing the strategic pass during the Atlanta campaign (illus. 30). Although expensive to produce, and therefore usually reserved for senior commanders, sun-printed maps could be quickly updated as operations unfolded, and sometimes several versions were made in a single day.

Mapping and surveying projects multiplied during the remainder of the nineteenth century, concurrent with colonial and imperial territorial ambitions. In the U.S., expansion westwards into Native American territories, and the ongoing Indian Wars, were supported by systematic photography by Timothy O'Sullivan, who had risen to prominence as one of Brady's freelancers during the Civil War. Between 1867 and 1869, O'Sullivan worked for the U.S. Geological Exploration of the Fortieth Parallel, a quasi-geographical study of Native American territories, sponsored by the U.S. War Department and the army's Chief of Engineers. U.S. expansion transformed landscapes and decimated peoples, although O'Sullivan's survey photographs have provided posterity with an invaluable historical record of the American West.

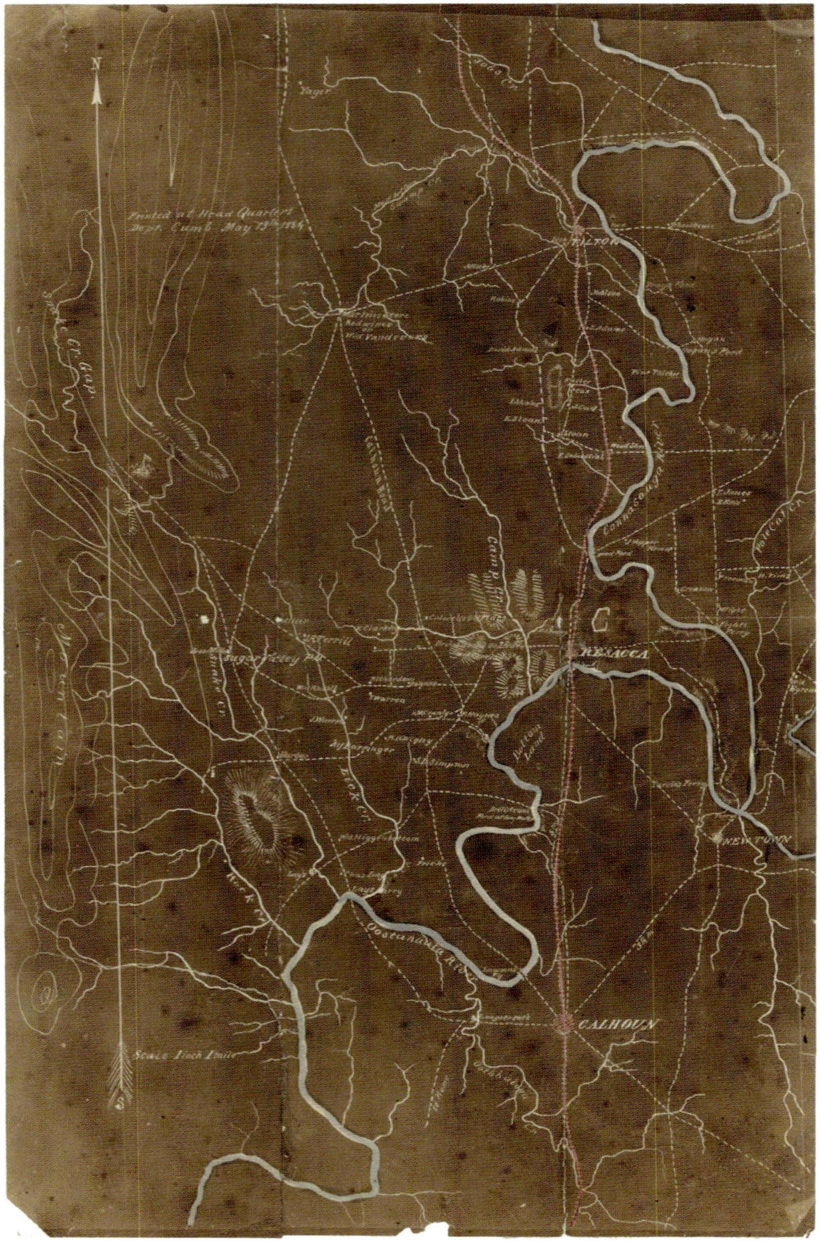
N
Printed at Head Quarters
Dept. Cumb. May 13th 1864
TILTON
RESACCA
Sugar Valley P.O.
NEWTOWN
Oostanaula River
CALHOUN
Scale 1 inch 1 mile
S

In the Middle East the Sinai Peninsula of Egypt was the subject of a British Ordnance Survey launched in 1868, with the strategic importance of the almost completed Suez Canal of particular interest.[15] Yet military photographic surveys were not only undertaken by Western powers. In 1861 the Egyptian army engineer Colonel Muhammad Sadiq was commissioned by the Wāli (Governer) of Egypt and Sudan, Sa'id Pasha, to survey Arabia between the Red Sea, the port of al-Wajh and Medina. Although photography was not officially part of this commission, Sadiq took a wet-plate camera along with his surveying tools and made images to supplement his detailed reports of the terrain and people of the region. After reporting to his military superiors, some of his images were published in the *Egyptian Military Gazette* in 1877. Sadiq made a further exploration of the region in 1880. As an Arab and a Muslim, he seems to have encountered little resistance to photographing holy sites, and made the first photographs of Medina and Mecca.[16] The Ottoman Army, meanwhile, introduced photographic training into its military schools, taught by Ali Sami Aközer, Servili Ahmed Emin and Ali Riza Bey, among others. An impressive collection of albums, accumulated by photographic enthusiast Sultan Abdulhamid II between 1880 and 1893, attests to the extensive photography of fortifications, weaponry, troops and naval vessels in the Empire.[17]

In Persia (now Iran), the photographer 'Abdollah Mirza Qajar surveyed the country's strategically important northern border regions and outlying dominions from 1883 to 1896. As Elahe Helbig has shown, photography was deployed by the Iranian authorities as a fundamental means of exerting control in a region under threat both from Russian and British colonial powers, and from unstable local rulers.[18] The *Albumkhaneh* (House of Albums) in the Golestan Palace in Tehran conserves hundreds of expeditionary photographs by 'Abdollah Mirza Qajar, documenting transport routes, communications and bridges, as well as people of political or military interest. Detailed captions alongside the mounted albumen prints relate potentially useful intelligence. *Panorama of Lasjard*, for example, is supplemented by information on the network of secret rooms and tunnels used by villagers as a refuge in times of danger (illus. 27).[19]

During the First World War, military photographic cartography assumed mammoth proportions. The shifting geography of trench networks and

30 United States Army Corps of Topographical Engineers, environs of Resaca, Georgia, 13 May 1864, map copy made using photosensitive paper.

repeated devastation of terrain by heavy artillery meant that maps soon became obsolete, and redrafting became a perpetual undertaking to keep pace with the changing topography. On the British side, Ordnance Survey staff and Royal Engineers mapped some 15,500 square kilometres (6,000 sq. mi.) of the Western Front, while the Germans set up the Kriegsvermessungswesen (Military Surveying Department) and printed over 250 million maps over the course of the war.[20] As the stalemate of trench warfare meant that advances were counted in yards, ever-more detailed maps were required. A section of a larger panoramic view made in 1915 by German forces shows the terrain of Ypres in Belgian West Flanders (illus. 31). The photograph was taken from an artillery observation post behind German lines in the Ypres area, and charts an area extending from Zillebeke to Bellewaarde Lake, including Menin Road and Hooge. Photogrammetric annotations that measure linear distances on the ground form a dispassionate counterpoint to the line of blasted trees that cross the horizon, while the copperplate script introduces an incongruous note of elegant civility to the scene. While ground-level mapping remained important throughout the First World War, the development of manned aircraft would introduce a new dimension to military surveying.

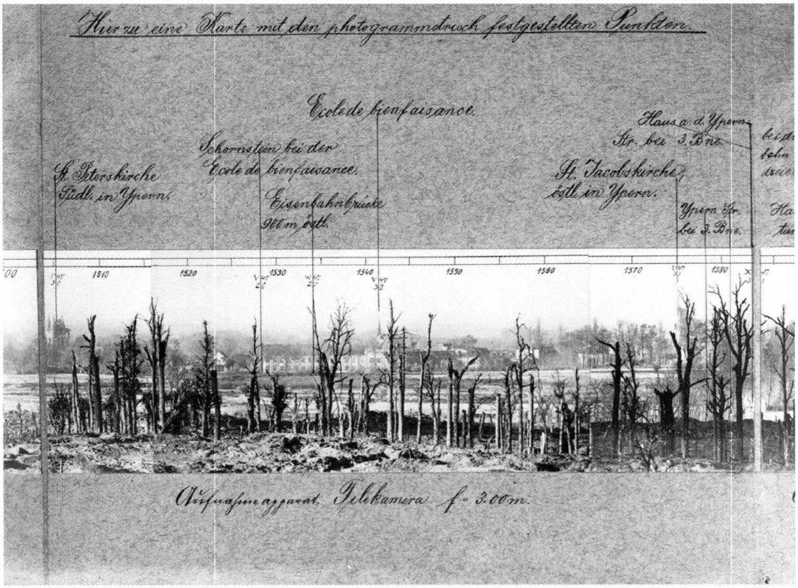

31 Photographer with the German armed forces, panoramic view of Ypres (detail), November 1915, silver gelatin prints, mounted and annotated.

Aerial Photo Reconnaissance

The military potential of aerial photographs had long been recognized, and kites, balloons and even pigeons were employed before manned, powered aircraft became available.[21] In 1859 Napoleon III offered the French balloonist and aerial photography pioneer Nadar (Gaspard-Félix Tournachon) 50,000 francs to make aerial views of the war against Austria in Italy. The commission was not undertaken, but Nadar began analysing aerial photos made from various angles, laying the foundations for stereophotogrammetry.[22] During the American Civil War, the Union Balloon Corps made at least one attempt to make photographs, but were foiled by high winds. In any case, the limitations of photographic technology at that time – lengthy exposure times and the need for a wet-plate darkroom to be incorporated into the balloon gondola – coupled with the time required for launches and descents, made such endeavours unviable.[23]

By the end of the century, however, technology had dramatically evolved. The U.S. Army incorporated photography into its Signal Corps, and kite and balloon photography was used above Puerto Rico in the Spanish–American War of 1898. Soldiers were trained in lab-work and photography at the Signal Corps NCO School at Fort Riley, Kansas, with one of the first instructors reputed to be Mathew Brady.[24] The British forces were conducting aerial observation in South Africa: one of the earliest extant military aerial photographs depicts the British camp's balloon base during the relief of Ladysmith in the Boer War, 1899.[25] A few years later, the German engineer Alfred Maul patented a system that could launch cameras – equipped with parachutes to ensure a soft landing – into the air with the aid of rockets.[26]

These innovations were superseded by the first successful manned powered flight in 1903, and the First World War became the first conflict in which reconnaissance by aerial photographers played a significant role. Continual dawn patrols undertook dangerous but vital surveillance of the ever-shifting territory from above.[27] The quality of cameras and lenses was crucial to effective photography, and both the Allies and Central Powers rushed to develop technology, with the British supplementing their inferior lenses with those captured from the Germans.[28] Equally important was protective cover in the skies. Most of the great aerial battles and campaigns of the conflict were concerned with attacking or defending photo reconnaissance aircraft, and the rise of the dashing fighter pilot – epitomized by German flying ace the 'Red Baron' (Manfred von Richthofen) – is one of the many unintended cultural consequences of military photography. Ironically, the photographers themselves remain unsung heroes of the First World War, despite experiencing some of the highest casualty rates.

32 Photographer with the German Luftwaffe, aerial view of Dalmuir, Clydebank, Scotland, 2 October 1939.

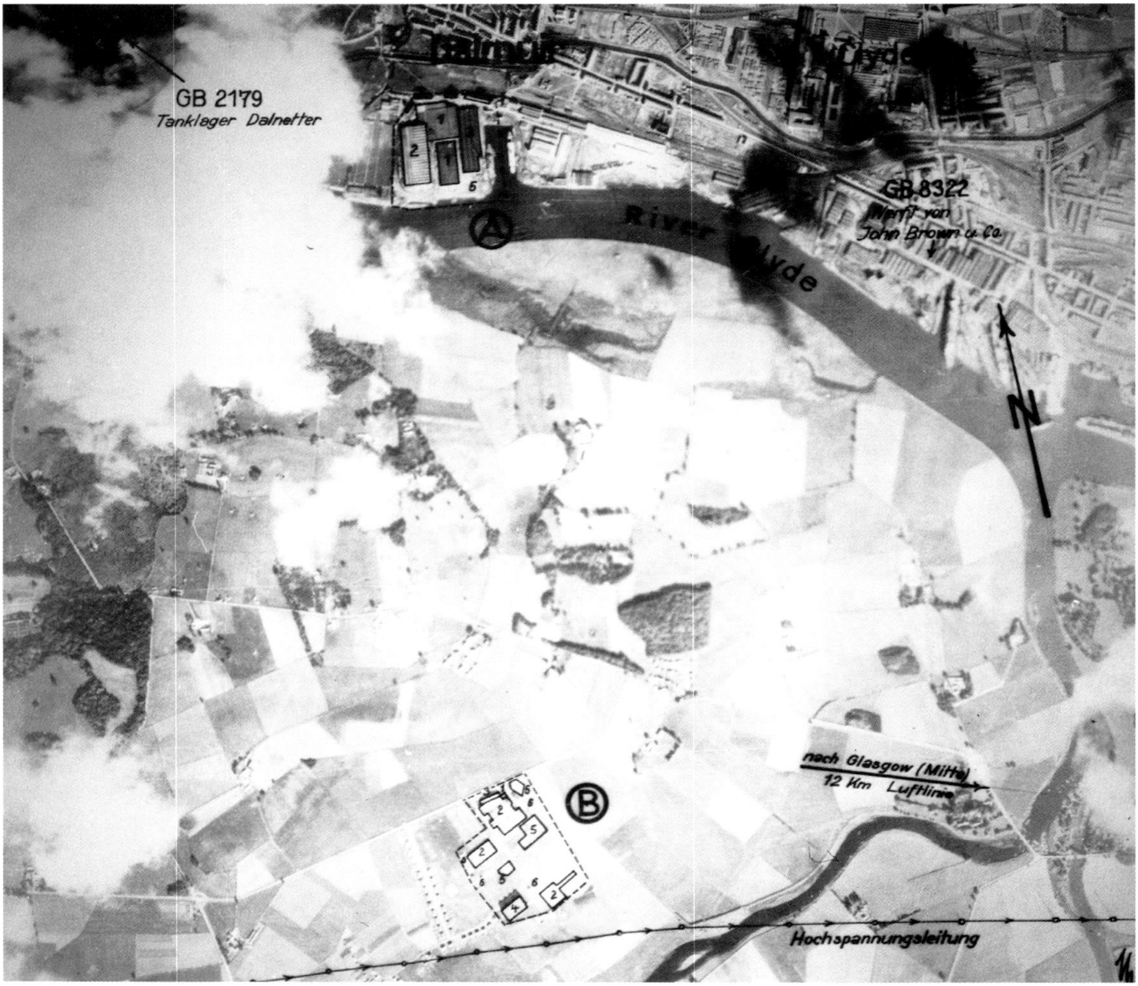

Reconnaissance by aircraft remained indispensable in the Second World War. Germany placed great importance on aerial intelligence, with General Werner von Fritsch, the chief of General Staff, declaring in 1938 that 'the nation with the best photo-reconnaissance will win the war.'[29] Hitler commissioned an extensive aerial survey of Europe, initially undertaken clandestinely by German military intelligence from civilian Lufthansa aircraft, and then during the war by the Luftwaffe air force. An image annotated on 2 October 1939 reconnoitres the strategic industrial area of Clydebank, to the west of Glasgow. The Royal Ordnance Factory, India Tyres and John Brown & Co. (one of the principal shipyards) have been marked with the letter 'A', indicating their interest as potential targets. (illus. 32). The area was subsequently bombed in March 1941, devastating not only the shipyards but munitions factories, oil storage tanks and – as collateral damage – almost 12,000 civilian houses. Yet although Germany possessed the most extensive aerial intelligence, poor interpretation and a lack of an integrated reconnaissance and interpretation unit ultimately limited its effective application.[30]

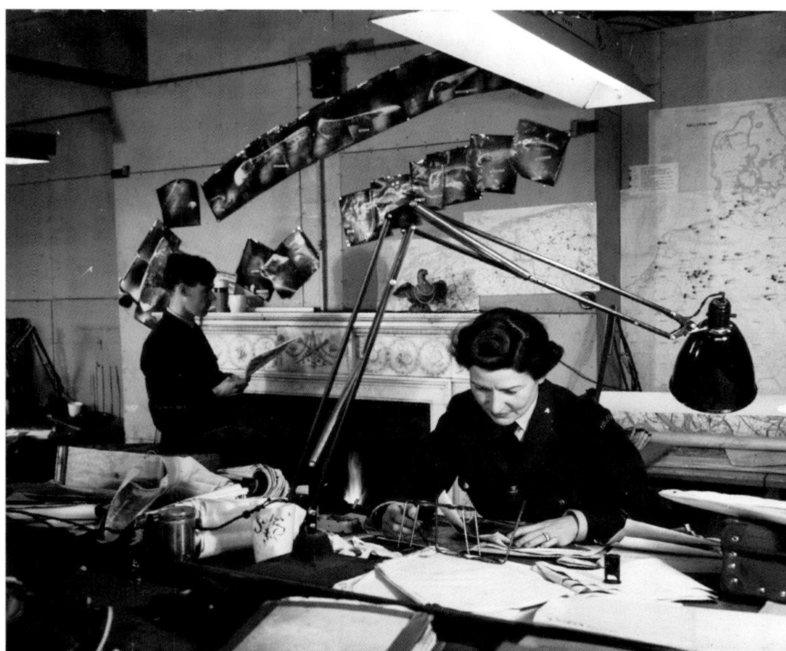

33 Photographer with the British Armed Forces, Joan Vyvyan Slade using a stereoscope to interpret aerial photographs, Photographic Interpretation Unit at RAF Medmenham, c. 1944, silver gelatin print.

By contrast, the Royal Air Force excelled at analysis. Highly skilled PIs (Photographic Interpreters) used overlapping 'mosaic maps' and, crucially, stereoscopes to make sense of raw data provided by aerial photographs (illus. 33). This complex work was undertaken in the unlikely location of a stately home in rural Buckinghamshire, which had been commandeered as RAF Medmenham. Women made up half the personnel, including Sarah Oliver, daughter of the Prime Minister, Winston Churchill, and Constance Babington Smith, who headed the Aircraft Section that made the crucial discovery of German V-1 'doodlebug' bombs at the Peenemünde weapons development site on the Baltic Coast.[31] Despite these decisive contributions, women PIs – in common with the rest of the Allied women's services – received only two-thirds of male pay rates.[32]

As the Second World War segued into the military stalemate of the Cold War, long distance 'remote sensing' assumed an even greater magnitude.[33] The critical role played by aerial reconnaissance was thrown into sharp relief when, for thirteen days in 1962, East and West teetered on the verge of nuclear conflict. The Cuban Missile Crisis was triggered by images of San Cristobal in western Cuba (illus. 34), taken by Lt. Col. Richard S. Heyser from a U.S. Air Force U-2 high-altitude spy plane. The Central Intelligence Agency's National Photographic Interpretation Center confirmed suspicions that the Soviet Union was constructing ballistic nuclear missile facilities on the island, just 145 kilometres (90 mi.) south of Florida.[34] The annotated photographs were circulated to the White House Executive Committee, prompting the decision to 'quarantine' military shipments to Cuba, and consequently escalating tension between the USSR and the U.S. to the brink of mutual annihilation. Eventually, President John F. Kennedy and the Soviet leader Nikita Khrushchev reached a secret agreement to bilaterally withdraw arms from a number of sites, thus averting a cataclysmic conclusion to the crisis. The aerial photographs of trees and buildings, impenetrable to the non-specialist, appear curiously banal and inconsequential given the fundamental role they played in the global history of warfare.

Cold War surveillance capabilities reached even greater heights, with satellites in space able to produce high-resolution images of distant territories. Aerial imagery from Zenit satellites, launched in 1962 under the cover of the

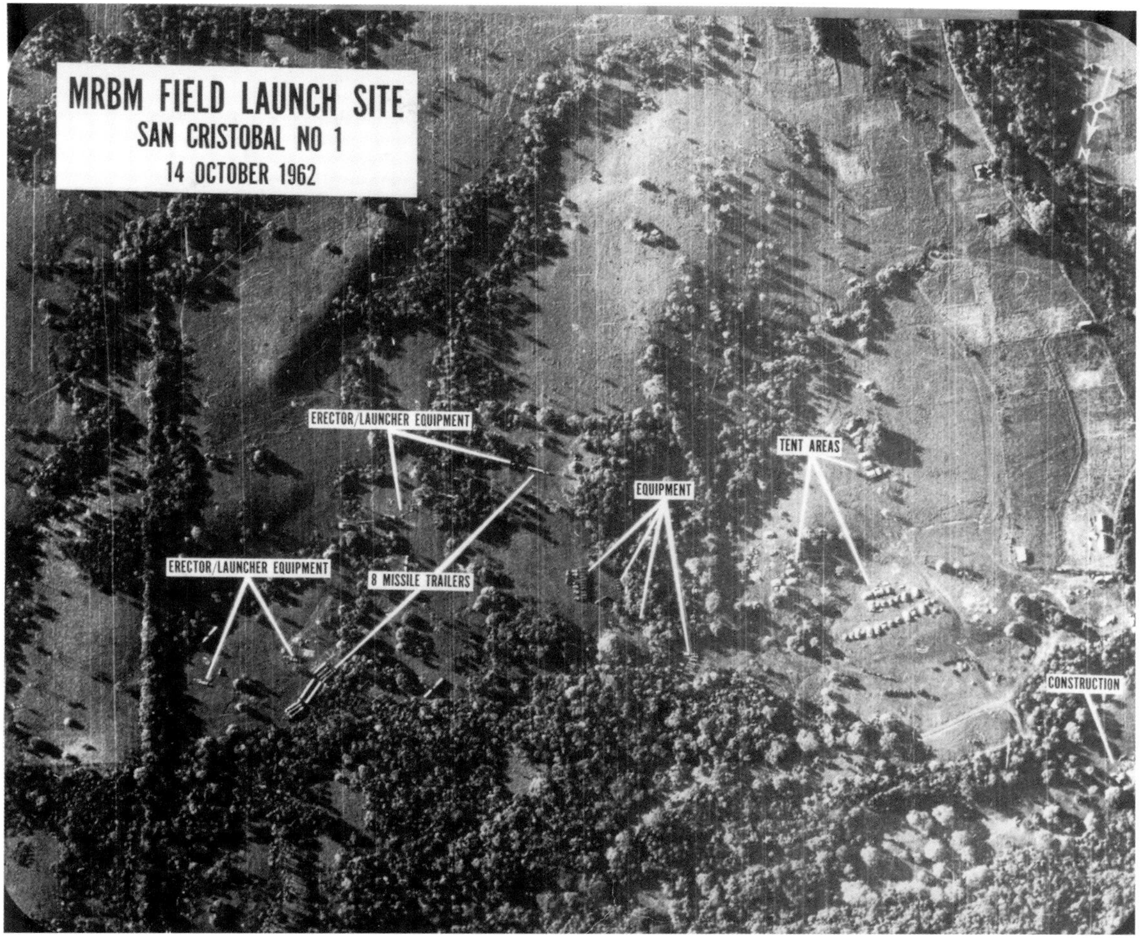

34 Lt. Col. Richard S. Heyser, U.S. Air Force, aerial view of missile base San Cristobal no. 1, Cuba, 14 October 1962.

scientific research programme Cosmos, provided a vital source of data for the exhaustive Soviet cartographical project carried out from the 1950s to the 1990s. The U.S., meanwhile, developed reconnaissance satellites that operated clandestinely or under the auspices of meteorological research, such as the Discoverer and Raincoat programmes of the late 1950s and early 1960s.[35] Geospatial intelligence, gathered from satellites, continues to inform national security units such as the National Geospatial-Intelligence Agency of the U.S. Department of Defense.

The advent of digital technologies has speeded live transmission of photographic imagery. As part of Operation Just Cause, the U.S. invasion of Panama (1989–90), infantrymen were accompanied by Pictorial Detachment soldiers equipped with still video cameras. Images were transmitted directly to the Pentagon, and for the first time live images from the battlefield were accessible to decision-makers hundreds of kilometres away.[36] Shortly after, Kodak's first digital still photo system – formed of a Nikon camera and a briefcase-sized hard-drive – was deployed in Kuwait and Saudi Arabia as part of Operation Desert Storm (1991) during the First Gulf War.[37]

Aerial reconnaissance remains crucial in the global War on Terror, and imaging systems now encompass live digital feeds from unmanned military drones that may be flown by operators sitting thousands of kilometres away. Their surveillance potential is vast: the Sierra Nevada Corporation's Gorgon Stare system, used by the U.S. military in Afghanistan in 2011, gathers visual data from multiple cameras to offer, according to its manufacturers, 'city-sized, 24/7 persistent surveillance'.[38] At the other end of the scale, developments in miniaturization may soon enable drones no bigger than an insect to fly within feet of an enemy.[39] The lethal capability and unmanned aspect of drones remains controversial. Proponents argue for greater accuracy of air strikes and fewer casualties of army personnel; detractors point to the legal and moral grey area of remote killing, and to the continuity of civilian casualties.[40] Whatever the implications, the military value of drone surveillance seems likely to remain a key part of the arsenal of the U.S. and its allies.

Contemporary Artists and Critiques of Military Vision

The potentially repressive use of the photographic gaze has been imaginatively interrogated by contemporary artists, sometimes subverting the very tools that are deployed by the military. Gaza-born Taysir Batniji documents structures in Palestine's West Bank with an apparently dispassionate precision that recalls the work of the Royal Engineers in nineteenth-century Afghanistan (see illus. 28). Presented as a typological survey (illus. 35), his series *Watchtowers* (2008) alludes to the approach of Bernd and Hilla Becher, the influential

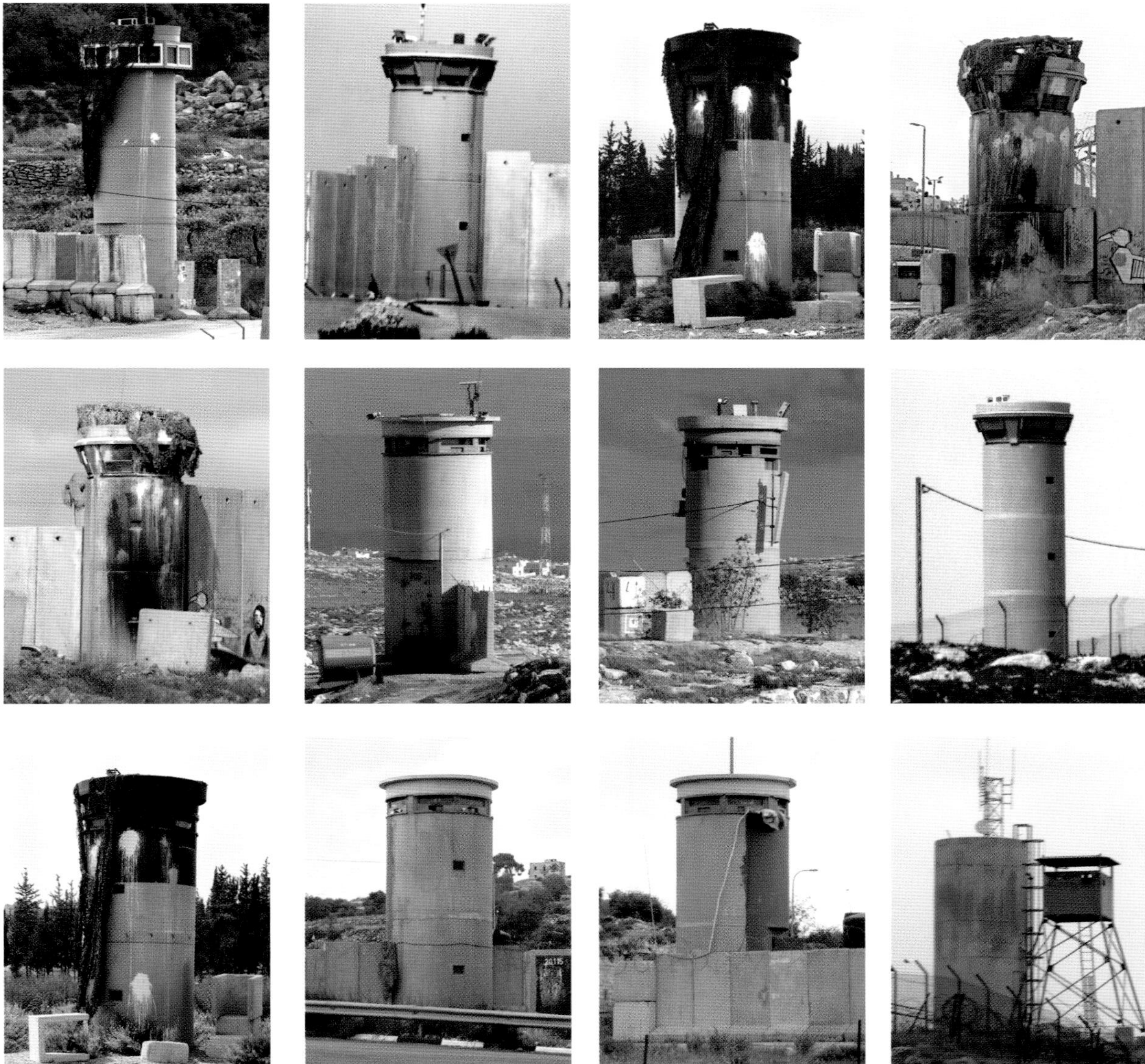

35 Taysir Batniji, *Watchtowers*, 2008, silver gelatin prints.

German modernist photographers who began working in the 1960s. While the Bechers celebrate the formal qualities of civilian structures such as water towers, gas tanks and pitheads, Batniji highlights the ubiquity of the repressive lookout tower: an everyday manifestation of occupation in the ongoing Palestine–Israeli conflict.

Iraq-born Jananne Al-Ani critiques the Western aerial gaze on the Middle East in her film *Shadow Sites II* (2011) (illus. 36). The title refers to a method of aerial archaeology – a by-product of military reconnaissance – whereby ancient constructions, virtually undetectable from the air, are revealed by their raking shadows at dawn and dusk.[41] Al-Ani draws on visual references including military and archaeological reconnaissance, as well as the Western media's reportage of the Desert Storm campaign. In such images, she asserts, 'the site of war [is] shown to be a desert, a place with no history and no population – an empty space, a blank canvas.'[42] Al-Ani's film was made using high-resolution photographs of the Jordanian desert shot from a tiny two-seater plane. Centuries of activity are revealed, from Bronze Age copper mines to Roman forts, First World War trenches to modern industrial farm

36 Jananne Al-Ani, still from *Shadow Sites II*, 2011, single-channel digital video (photography Adrian Warren).

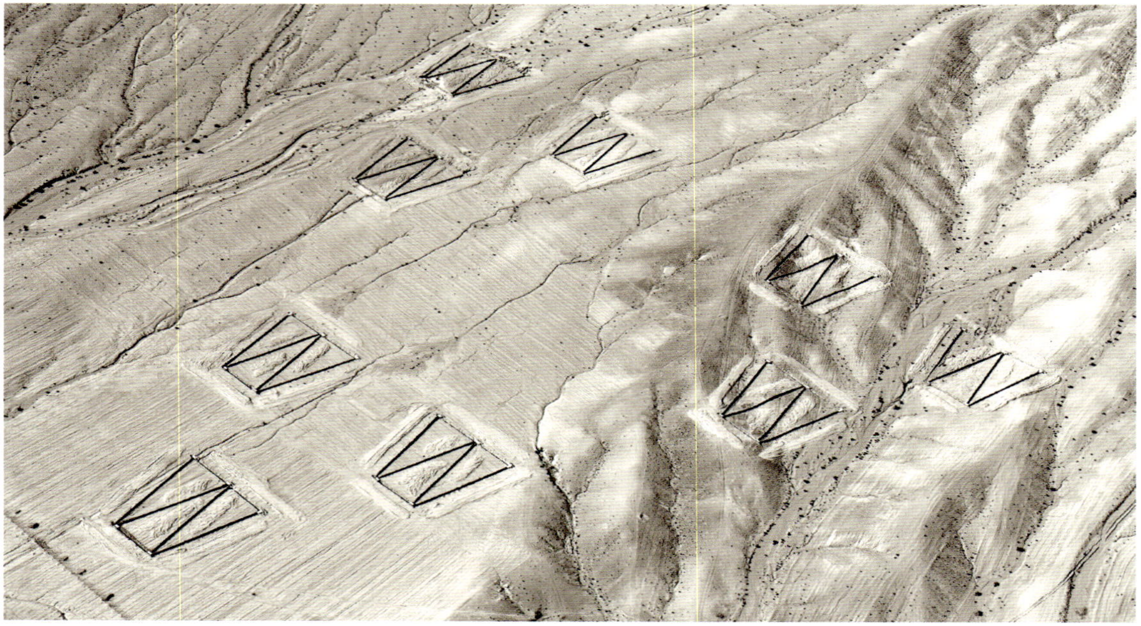

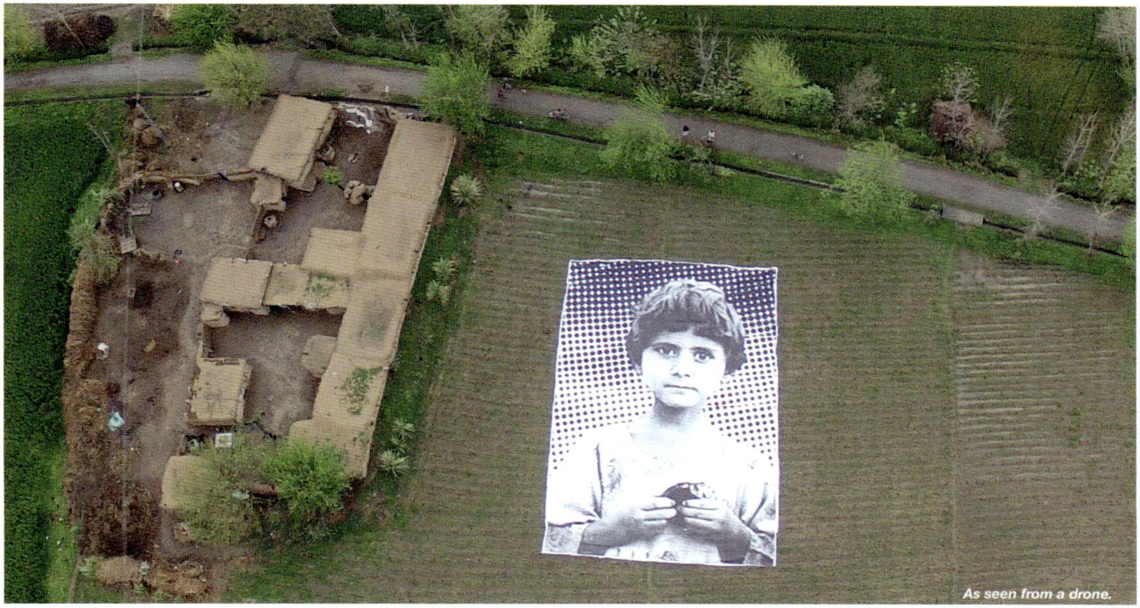

37 Not a Bug Splat, aerial view of a photographic print installed in Khyber Pakhtunkhwa, Pakistan, 2013.

complexes: 'When viewed from above,' the artist states, 'the landscape itself acts as a photographic plate, where a latent image [...] is periodically revealed as the sun rises and sets over a site'.[43] Far from being empty, the landscapes of *Shadow Sites II* are replete with signs of historical and ongoing human construction and activity.

Present-day modes of warfare, characterized by remote, covert and digital operations, have proved resistant to war photography in the traditional photojournalistic mode, which relies on images of human drama and combat action.[44] In any case, the practice of 'embedding' journalists with U.S. military units since the 2003 Invasion of Iraq has severely limited opportunities for critical viewpoints. Artist-activists have offered alternative strategies, becoming increasingly creative in their attempts to make new modes of warfare visible. The Pakistani-U.S.-French collective Not a Bug Splat takes its name from U.S. military slang used by Predator drone operators, who allegedly refer to remote human kills viewed on screen as 'bug splats'.[45] The artist collective made spectacular use of a photographic image of a young girl taken by Noor Behram, a photojournalist from

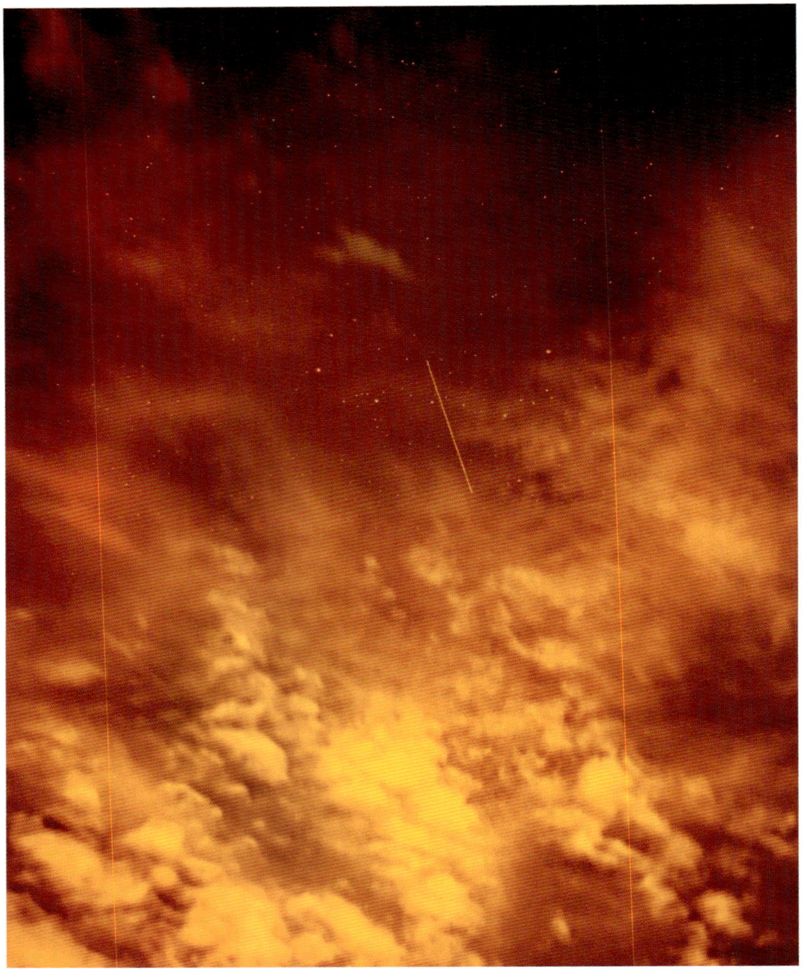

38 Trevor Paglen, KEYHOLE 12-3 (IMPROVED CRYSTAL) *Optical Reconnaissance Satellite Near Scorpio* (USA 129), 2007, C-type print.

Waziristan.[46] A giant print featuring the girl, claimed to have been orphaned following drone strikes, was installed in a field next to a village in Khyber Pakhtunkhwa, the scene of a U.S. attack in 2013 (illus. 37). The artist collective hoped that this image, seen on camera thousands of kilometres away, might lead drone operators to empathize with victims and question the ethics of their task. While the effects on individual operators are unknown, the collective was certainly successful in garnering international news coverage, helping to keep the issue of civilian victims of drone strikes on the agenda for public debate.[47]

The U.S. artist and political geographer Trevor Paglen has produced some of the most sustained and compelling work on military surveillance. For *The Other Night Sky* (2007–11) Paglen worked in collaboration with amateur astronomers, using astrophotography to map and depict 'black' (clandestine) U.S. satellites and spacecraft.[48] Working from the desert valley beneath the Sierra Nevada mountains, Paglen uses a sophisticated combination of telescopic, digital and large-format film photography. The trajectories of the satellites often appear in his images as small streaks across the skies, such as *Keyhole 12-3*, shown near the constellation of Scorpio (illus. 38). Paglen's uneasily beautiful images invite prolonged reflection, and are equally at home in an art museum as the pages of a technology magazine.[49] Like other artists who interrogate surveillance and remote warfare, Paglen believes that making covert operations visible is the first step to public debate in a democratic society.

Corporate Innovation, Penetrative Vision

Commercial interests and military needs are closely intertwined, and civilian corporations have played a decisive role in the development of technologies for war. Large photographic companies, such as Kodak in the U.S., Agfa, Leica and Zeiss in Germany, and Canon, KonicaMinolta, Tokyo Optical, Fujifilm and Nikon in Japan, developed symbiotic relationships with state and armed forces, with ample military funding supporting the development of new products and processes.[50] In the 1930s, for instance, the Japanese government supported national innovations in high-speed, X-ray and infrared film, as well as underwater cameras.[51] Kodak, meanwhile, became the U.S. Army's preferred supplier of photographic materials. By the end of the First World War, 38 per cent of Kodak's contracts were with the U.S. military; during the Second World War virtually all production was directed to war use.[52] Kodak's 'Serving Human Progress' marketing campaign went so far as to claim that its photographic innovation was modernizing the U.S. armed forces.[53] One of its advertisements boasted that 'more than a hundred war products [are] now made of material

PHOTOGRAPHY'S "SECOND FRONT"

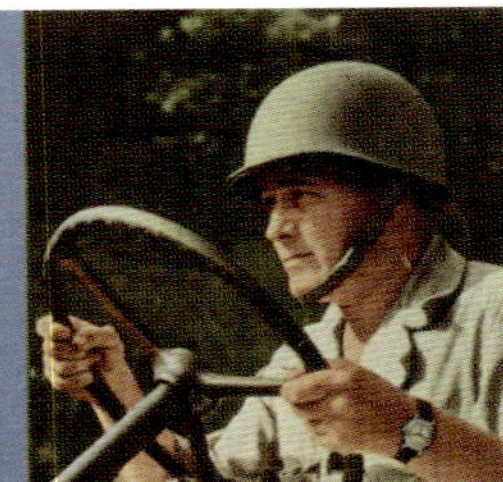

He controls the Jeep with a Tenite steering wheel—strong, tough, and able to stand all climates. Your own car probably has a Tenite steering wheel, instrument panel, accessories.

More than a hundred war products now made of material developed for a better Kodak Film

FILM BASE IS A PLASTIC—one of the earliest. To make a better film, Kodak long ago began producing from cotton linters a "miracle material": cellulose acetate.

In the form of TENITE—made by Tennessee Eastman Corporation, a Kodak subsidiary—this plastic is tough as a steer's horn and lighter than wood. It can be molded under heat or pressure, or "machined" like lumber or metal. It can be clear transparent, or in an unlimited range of colors.

Tenite is molded into finished products at the fastest rate ever reached with plastics. It led to a minor "industrial revolution" before the war or wartime shortages were dreamed of . . .

Now it has more than a hundred war applications —not as a substitute, but as a superior material. As an extra advantage, it does supplant other "critical" materials.

A few war uses are illustrated . . . In a sense, they all started with photography—the ever-growing need for finer film . . . Eastman Kodak Company, Rochester, N. Y.

REMEMBER TORPEDO SQUADRON 8? *. . . how, knowing exactly what the odds against them were, this heroic band of 30 Navy fliers drove unswervingly into the massed fire of the Japanese fleet off Midway? And only one man survived? A stern example to us at home.* **BUY MORE WAR BONDS.**

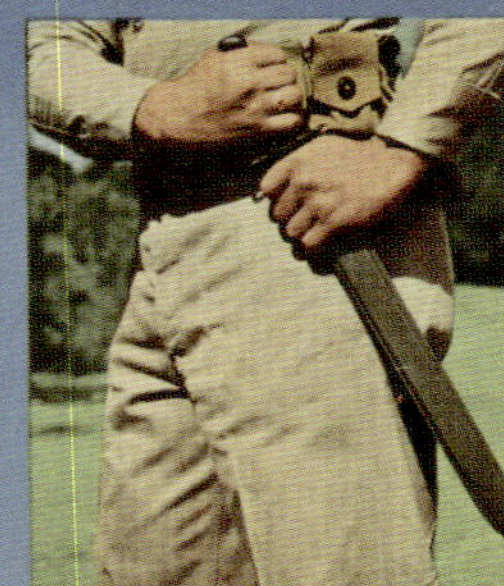

His bayonet scabbard is Tenite—lighter, tougher, more easily cleaned . . . Cost is little more than half that of scabbards made with earlier materials.

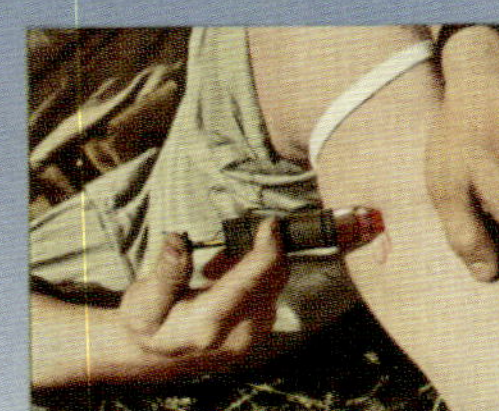

Snake-bite kit supplied our troops by the Army Medical Corps includes vacuum pump—molded of Tenite—for extracting snake venom.

Doubles for brass—Molded of Tenite, this bugle once more raises the question, "What will plastics do next?" Before acceptance by the Army, it won the most critical ears by its tone and range.

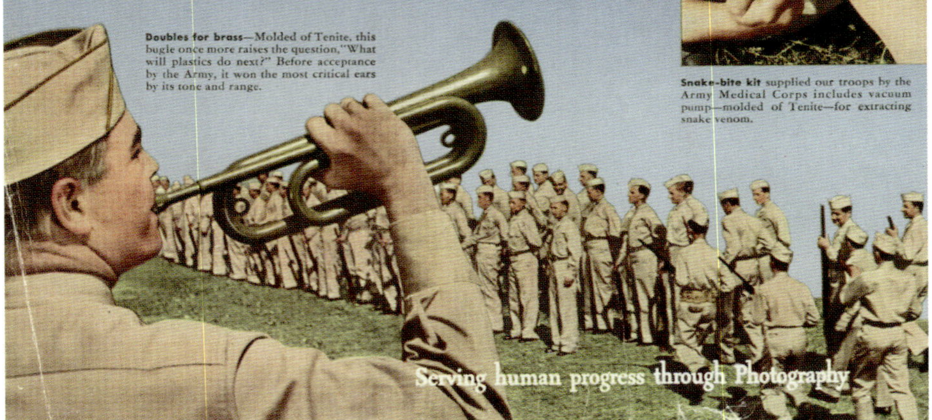

developed for a better Kodak film' (illus. 39). Tenite, a plastic developed from cellulose acetate for film base, was apparently used to make steering wheels for jeeps and bayonet scabbards; it even replaced brass used for that most patriotic of symbols, the army bugle. Whether corporate innovation was really leading the way is a matter of conjecture, and the nature of photographic-military-industrial relationships is far from clear-cut.

Photographic and optical technologies have certainly enhanced military capabilities, sometimes surpassing the limitations of the human eye and the visible spectrum. The conquest of darkness has been an ongoing military preoccupation, and explosive 'star shells' and 'light balls' providing illumination rather than arson were employed as early as the seventeenth century.[54] In the Second World War the U.S. Army Air Force sought a more efficient alternative to power bombs to provide high-intensity light to aid night-time observation and targeting. The private photographic company EG&G, co-founded by scientists Harold Edgerton, Kenneth Germeshausen and Herbert Grier under the auspices of the Massachusetts Institute of Technology, was commissioned to develop electronically generated strobe flash technology that could be carried by aircraft.[55] The system required a 76-centimetre (30-in.) quartz flash tube and banks of capacitors, weighing up to 227 kilograms (500 lb) each, to power a split-second flash of light capable of illuminating a target from the height of 1.6 kilometres (1 mi.). Integrated with radar and a synchronously timed camera, the strobe flash system proved invaluable on the night of 5 June 1944, when photographs confirmed that the D-Day invasion of Normandy would take the Axis forces by complete surprise.[56]

Infrared photography, sensitive to long-wave light, also enables penetrative sight beyond human vision. Following the publication of experiments by U.S. inventor Robert W. Wood in 1910, infrared's potential for haze penetration in aerial reconnaissance was developed during the First World War. By the 1930s, dye-sensitized silver halide (black-and-white) film was made commercially available by international manufacturers including Agfa, Kodak and Ilford.[57] 'False colour' infrared films, such as Kodak's groundbreaking Aerographic and Aerochrome, were developed for the U.S. Air Force in the Second World War.[58] The salient characteristic of infrared film is its capacity to distinguish between foliage and camouflage, making

39 Kodak magazine advertisement from the 'Serving Human Progress' campaign, c. 1941.

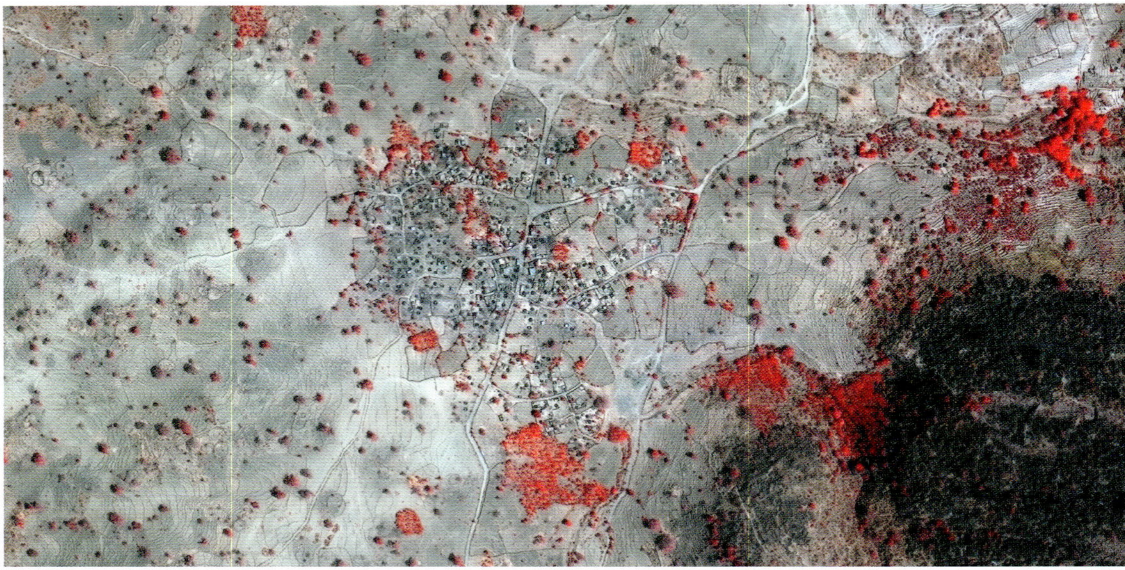

40 DigitalGlobe, aerial view of Darfur used in the Amnesty International campaign 'Decode Darfur', *c.* 2016, false colour infrared image.

it possible to discern troops in rural or heavily forested areas. Vegetation appears as glowing white areas on black-and-white prints; 'false colour' film translates the chlorophyll of green trees into vivid pinks and reds. The technology became particularly valuable in conflicts in Korea, Vietnam and Latin America, playing a decisive role in the defeat of Ernesto 'Che' Guevara's guerrilla campaign in Bolivia in 1967.[59] More recently, the ongoing 'War on Drugs' has seen the use of false colour aerial and satellite imagery to identify illegal coca plantations in Colombia.[60]

Such imaging is not limited to the military. Present-day private satellite companies such as DigitalGlobe have made some data commercially available, and the humanitarian organizations Human Rights Watch and UNOSAT (United Nations Operational Satellite Applications Programme) have acquired satellite images to assess the impact of hostilities on civilian zones in Syria and Gaza respectively. Amnesty International has made productive use of false colour infrared images made during the War in Darfur (2003 and ongoing), which unequivocally show scarlet areas of healthy vegetation that may be compared to grey areas of burned villages (illus. 40). Such imagery has the potential to function as evidence of ongoing conflict and human rights abuses in the face of denials by the Sudanese government.[61] Mobilized in this way, claims photography scholar Wilco Versteeg, machine-made photography may fulfil 'the role of witness and public defender', a role once occupied by the archetypal war photographer in the mould of Robert Capa or Philip Jones Griffiths.[62] Nonetheless, there are limitations to this potential: military agencies may restrict satellite data either by buying all commercially available images, or by imposing a resolution cap to degrade images accessed by outside parties.

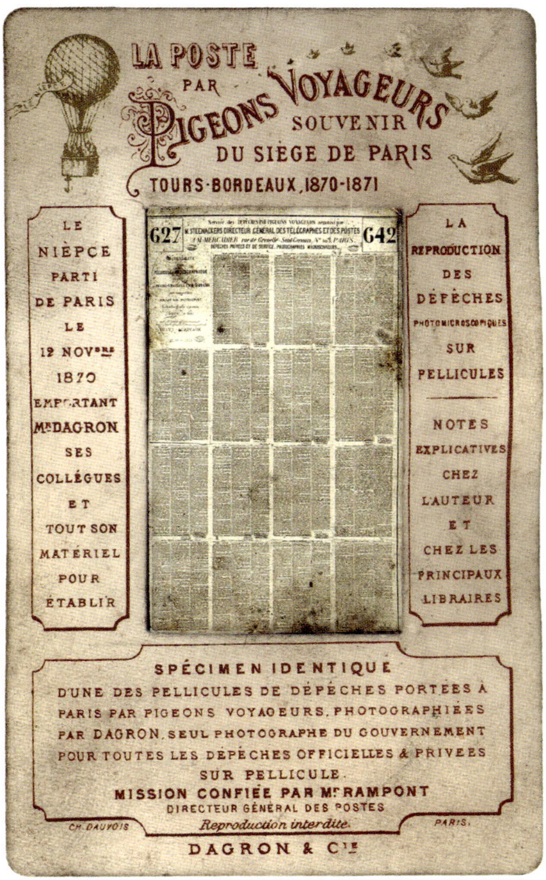

41 Dagron & Cie, 'Post by Travelling Pigeons: Souvenir of the Siege of Paris', c. 1871, card incorporating microfilm pellicle.

Military Communications

An overlooked but vital aspect of wartime photography is its role in communications, serving as a medium to replicate and disseminate information. Photography's usefulness in producing and reproducing military maps in the field has already been discussed, but a further application is the miniaturization and transportation of information. Microfilm, initially developed in 1839 by British inventor John Benjamin Dancer, was patented in France in 1859 by René Dagron, who correctly foresaw its huge potential for copying and miniaturizing documents.[63] The Franco-Prussian War of 1870–71 provided a striking instance of the usefulness of wartime microphotography, when carrier pigeons were used to carry microphotographs on collodion pellicle (a thin membrane of emulsion) across German lines between besieged Paris and the provinces.[64] A single pigeon could carry the equivalent of five hundred pages of official orders, fitted into a quill holder and attached to its tail feather. Upon receipt, the pellicle would be projected and transcribed. After the French defeat, the entrepreneurial Dagron obtained permission to reproduce despatches on pellicles, which he sold as card-mounted souvenirs (illus. 41). The accompanying text trumpets Dagron's role as 'the only government photographer for all official and private despatches on pellicle'.[65]

Microfilm remained a crucial means of wartime communication in the twentieth century. In the Second World War Kodak created a service called 'V-mail', adopted by the U.S. government as a practical solution to maintaining correspondence between home and frontline. Letter writing was considered essential to maintaining soldiers' morale, yet the continual shipment of millions of letters was a vast operation. The V-mail system entailed micro-photographing original letters, transporting the lightweight microfilm to the frontlines, and then printing out paper copies for recipients. Not all correspondents were content to have their letters mediated in such an impersonal way, however. As *Army and Navy* magazine reported on 14 February 1944,

> The patience of Army postal officers was at an end. They issued a stern edict: after St Valentine's Day, imprints of lipstick will no longer be tolerated on V-mail. Explained Major Kenneth H. Donnelly, postal officer of the Sixth Service Command: lipstick smears when it passes through V-mail photographic equipment, ruins the letter that bears it, and others that follow. The automatic feeder must be stopped and cleaned after every passing of 'the scourge'.

Servicemen complained that V-mails (and Airgraphs, the British version) were but poor imitations compared to physical letters bearing the unmistakeable imprint of their loved ones, but the system provided an invaluable chain of personal communication before the era of email.[66]

Stopping Cannonballs: Weapons Testing and Assessment

Photography's capacity for distanced viewing has long made it a valuable tool for analysing weapons in relative safety. During the American Civil War, the ever-eloquent commentator Oliver Wendell Holmes reported that

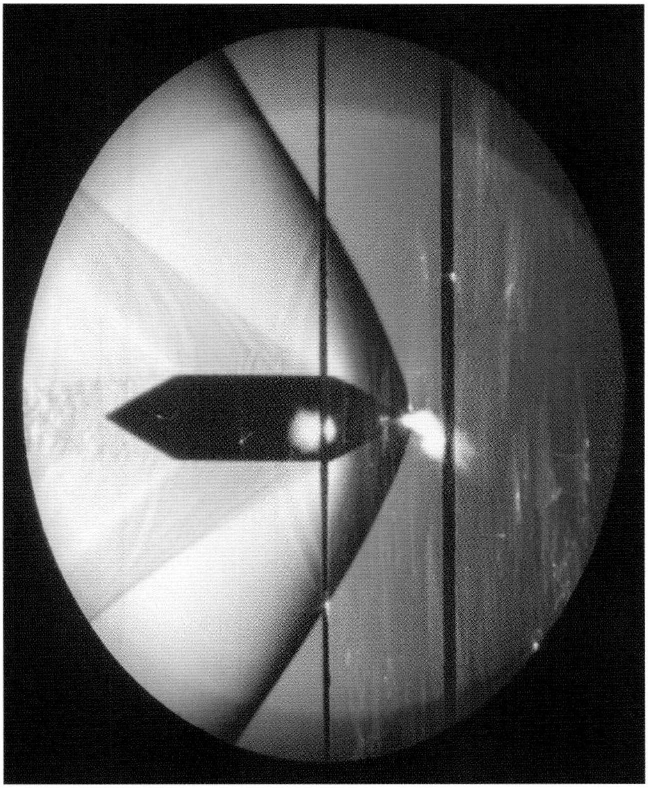

42 Ernst Mach, 'Double-pointed Brass Projectile, 530 m/sec., G. Werndl-Rifle', 11 September 1890, image from black-and-white glass negative.

> experiments upon the walls of forts with shot and shell, the damage done a vessel, the respective merits of military inventions, all these are recorded by the art [of photography] with more promptitude and fidelity than tongue or pen can achieve. It is even said that the passage of a cannonball through the air can be photographed – that the double-barreled lens of the sun battery shoots, as it were, the ferrigunous [*sic*] sphere as it flies.[67]

In reality, photographic technology was not sufficiently evolved to permit the capture of fast-moving objects; cannonballs, Holmes observed, 'unlike model little boys', were 'heard and not seen'. Nonetheless, military departments, commercial businesses, scientists and entrepreneurs began developing photography's capacity to function as an analytical tool of assessment. In 1866 the British Royal Artillery at Woolwich was attempting to record projectiles in flight, but German experiments soon led the field. From 1875 onwards, the German arms manufacturer Krupp began recording firing tests, both for diagnostic and marketing purposes.[68] While the effects of projectiles on their targets were recorded without difficulty, capturing the projectiles in flight was another matter. In the 1880s Moravian physicist Ernst Mach, along

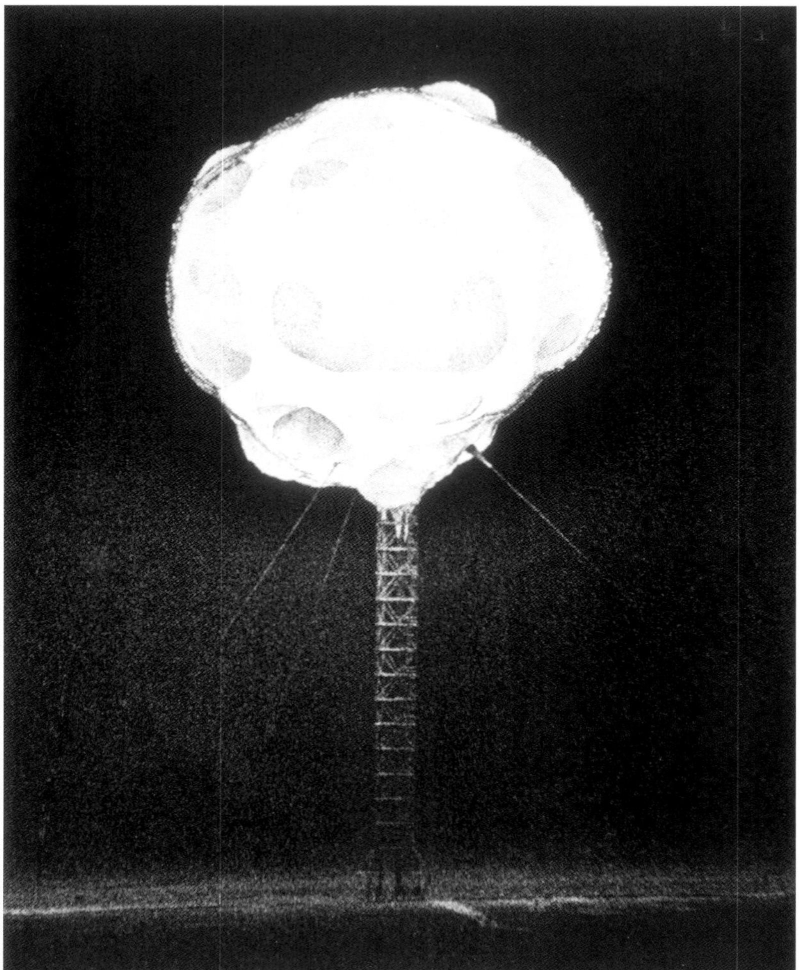

43 Photographer working for EG&G, record of an atomic explosion nuclear test made by a Rapatronic camera, Nevada Proving Grounds, *c.* 1952, black and white photograph.

with other innovators, experimented with electric spark technology that could be synchronized with fast-moving objects crossing darkened spaces. Drawn into a controversy over the use of illegal explosive bullets during the Franco-Prussian war (1870–71), Mach successfully photographed bullets fired by a pistol. In 1886 he was the first to record the shock waves from a bullet travelling faster than the speed of sound (illus 42).[69] Meanwhile, Prussian inventor Ottomar Anschütz experimented with linked cameras and focal-plane shutters to make instantaneous exposures fast enough to capture a moving subject. In 1882 he trialled his system on German troops during their field exercises, and in 1888 made a daylight photograph of a flying cannonshell, using focal plane shutters operating at 76 millionths of a second, at the Krupp works at Buckau-Magdeburg.[70] Holmes's prediction of being able to capture 'a ferriginous sphere as it flies' had become a reality.

Experiments in nuclear fission during the Second World War and the Cold War required even more sophisticated recording techniques. Robert Capa may have advised that 'if your pictures aren't good enough, you're not close enough', but the dangers of atomic weapons made it prudent to observe from a safe distance.[71] Throughout the 1950s and '60s, the EG&G company photographed nuclear tests using their specially developed Rapatronic camera that could record the explosions in exposures measured in nanoseconds. A series of 'fireball' images (illus. 43), made at the Nevada Proving Grounds for the Atomic Energy Commission, show the first microseconds of an atomic explosion. A photograph made in a 1/100,000,000-of-a-second exposure, taken from 11 kilometres (7 mi.) away with a lens 3 metres (10 ft) long, records a steel tower before vaporization in the intensity of the blast.

Many of the images produced by the Manhattan Project and U.S. Atomic Energy Commission remained secret, but some of the so-called 'mushroom cloud' images were publicly released and widely disseminated in magazines such as *Life*. As shown in Wayne Miller's photograph of children viewing an image of an atomic explosion on display at the 1955 exhibition *The Family of Man* (see illus. 3), such images had a propaganda role, functioning as icons of U.S. power that inspired both awe and fear. But as Kevin Hamilton and Ned O'Gorman have shown, EG&G's role went beyond merely documenting atomic tests.[72] At the request of the Manhattan Project, Herbert Grier solved the problem of how to actually detonate the bombs in mid-air, without reliance on cumbersome power units. EG&G's Raytheon factory in Boston, which had been producing capacitors for aerial night-time photography, quietly switched to developing firing-and-timing mechanisms for atomic bombs. The final Grier-designed device was used to detonate the Fat Man bomb over Nagasaki on 9 August 1945: perhaps the most spectacular instance of photography's contribution to military destruction.

The Military Medical Gaze

The rehabilitation of injured soldiers – whether prompted by humanitarian impulses to relieve suffering, or military pragmatism to maintain the fighting machine – has long been assisted by photographic technology. Images initially documented procedures such as operations, treatment and recuperation. One of the earliest examples, a daguerreotype from the Mexican–American War (1846–8), records an amputation undertaken by Dr Pedro Vander Linden, Inspector General of Mexico's Military Medical Corps. Made after the Battle of Cerro Gordo in Veracruz on 18 April 1847, the image shows the surgeon holding the amputated leg of the apparently semi-conscious sergeant Antonio Bustas: a scene which was almost certainly re-staged for the camera.[73] During the American Civil War, Federal medical authorities made more systematic photographic records of wounded soldiers. These images are a curious hybrid of aesthetics, inflecting the genteel parlour portrait with a scrutinizing medical gaze. A photograph made at Harewood Hospital, Washington, DC, under the direction of military surgeon Reed B. Bontecou, depicts Ludwig Kohn, a private who suffered a gunshot wound in the Battle of Gettysburg, 1 July 1863 (illus. 44).[74] Shirtless but wearing a military cap, Kohn is depicted in profile, the better to examine the angle at which the bullet passed through his chest. Now almost healed, his body gives little indication of its trauma, and supplementary text and a hand-drawn red arrow are obliged to convey the soldier's journey from 'gangrenous' wounds to his current 'very good constitutional state'. Bontecou compiled 570 *cartes de visite* of patients to produce a comprehensive medical album and enlarged a selection of them to aid in teaching.[75] Some of the images were later translated into engravings and reproduced in the six-volume

Medical and Surgical History of the War of the Rebellion, published between 1870 and 1888.

More sophisticated investigative photographic imaging came to the fore at the end of the century. The X-ray, discovered in 1895 by the German scientist Wilhelm Conrad Röntgen, was first used on soldiers returning from Italy's Ethiopian campaign, then closer to the front in a German Red Cross hospital in the Greco–Turkish War of 1897. The U.S. army equipped three hospital ships with X-ray facilities during the Spanish–American War of 1898,

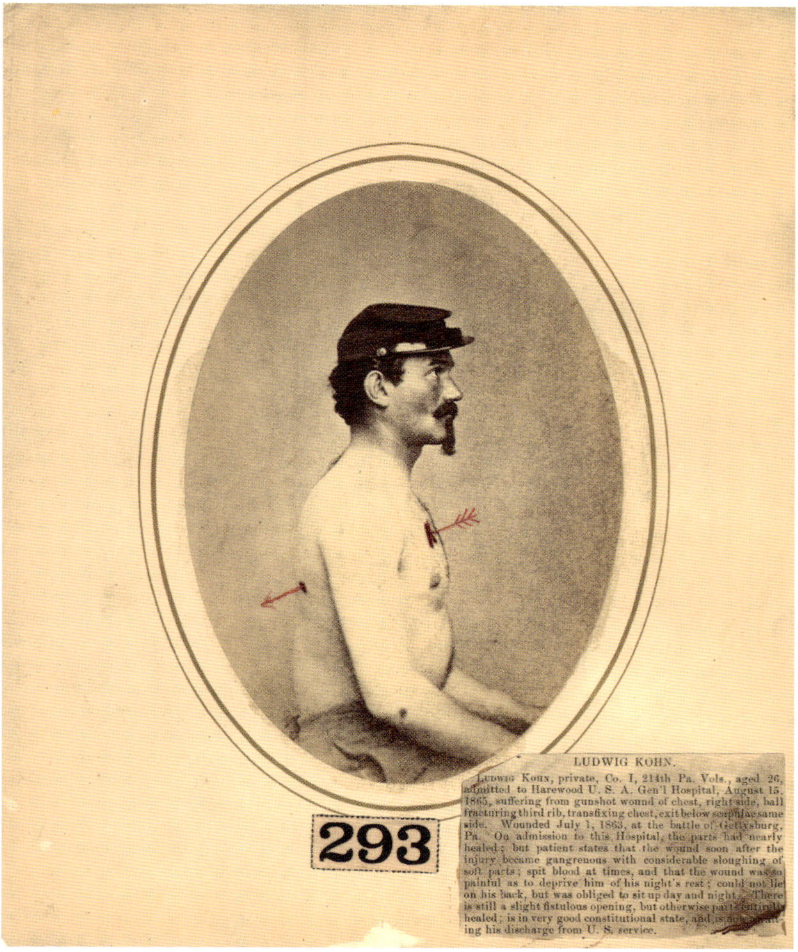

44 Reed B. Bontecou (attrib.), Private Ludwig Kohn exhibiting gunshot wound, Harewood Hospital, Washington, DC, c. 1863–5, mounted albumen print with annotations.

although most wounded soldiers were radiographed upon returning home.[76]

Women were effective practitioners in this field. Elizabeth Fleischman, for example, established an X-ray laboratory in San Francisco and was paid by the Surgeon General to provide services to the army. Her posteroanterior chest radiograph (illus. 45), made at the Oresidio Hospital in San Francisco, shows a Mauser bullet lodged in the chest of a soldier wounded in the Philippines during the Spanish–American War (1898). Fleischman's work was highly praised and her images were used to illustrate Capt. William C. Borden's landmark text *The Use of the Röntgen Ray by the Medical Department of the United States Army in the War with Spain, 1898*. Working continuously without radiation protection, however, Fleischman developed cancer and died in 1905 at the age of 38.[77] Women continued to be prominent radiologists in the First World War. The Polish-born scientist Marie Skłodowska Curie set up some two hundred X-ray stations in France and Belgium. As servicemen were continually being transferred for combat duties, Curie trained a permanent team of women to operate radiography equipment and helped establish a department of radiology at the Nurses' School of the Edith Cavell Hospital. In 1921 Curie published *La Radiologie et La Guerre* (Radiology and War), an illustrated publication that described ways of localizing bullets and shrapnel.[78]

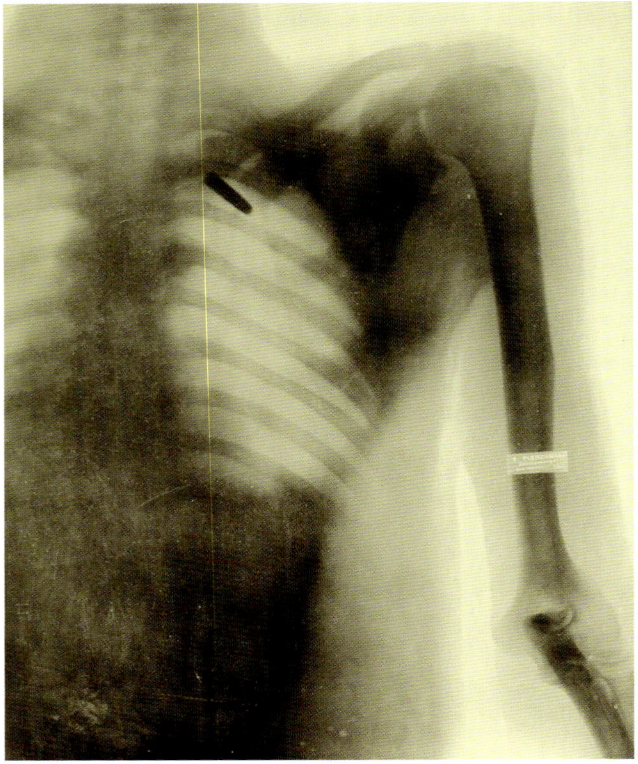

45 Elizabeth Fleischman, a Mauser bullet lodged in the chest of a soldier wounded during the Spanish–American War, 1898, radiograph.

Among the many medical photographers who have served in wartime, the trajectory of British practitioner Percy Hennell is especially noteworthy.[79] Originally trained as an artist, Hennell found more prosaic employment in the Colour Photographic Department of The Metal Box Company, where he was responsible for the design of tins for biscuits and confectionary. By 1939 Hennell had developed a technique using synchronous tri-chromatic filters

46 Percy Hennell, post-operative image of a male patient with injury to the nose and left eye, c. 1940–44, colour tri-chromatic print.

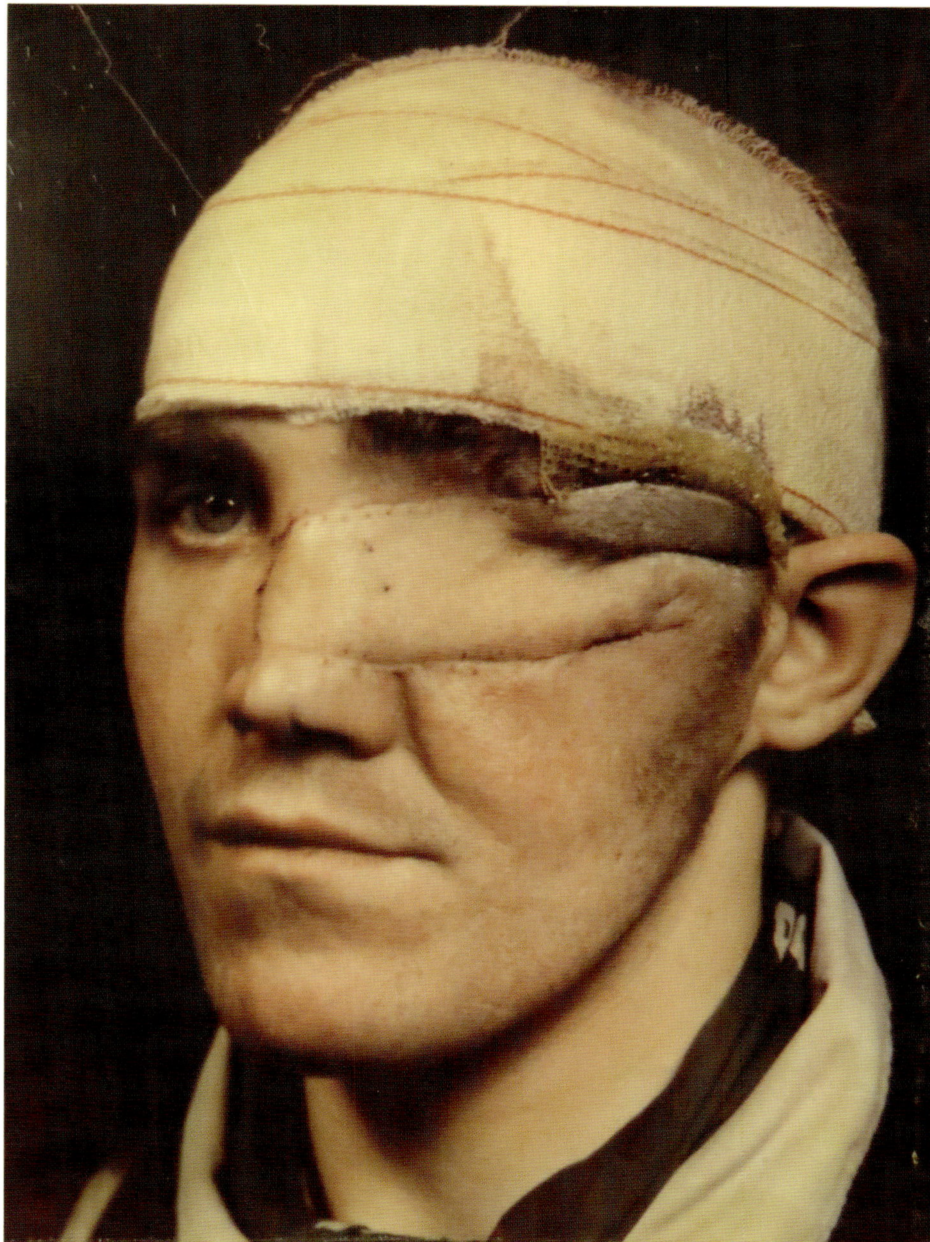

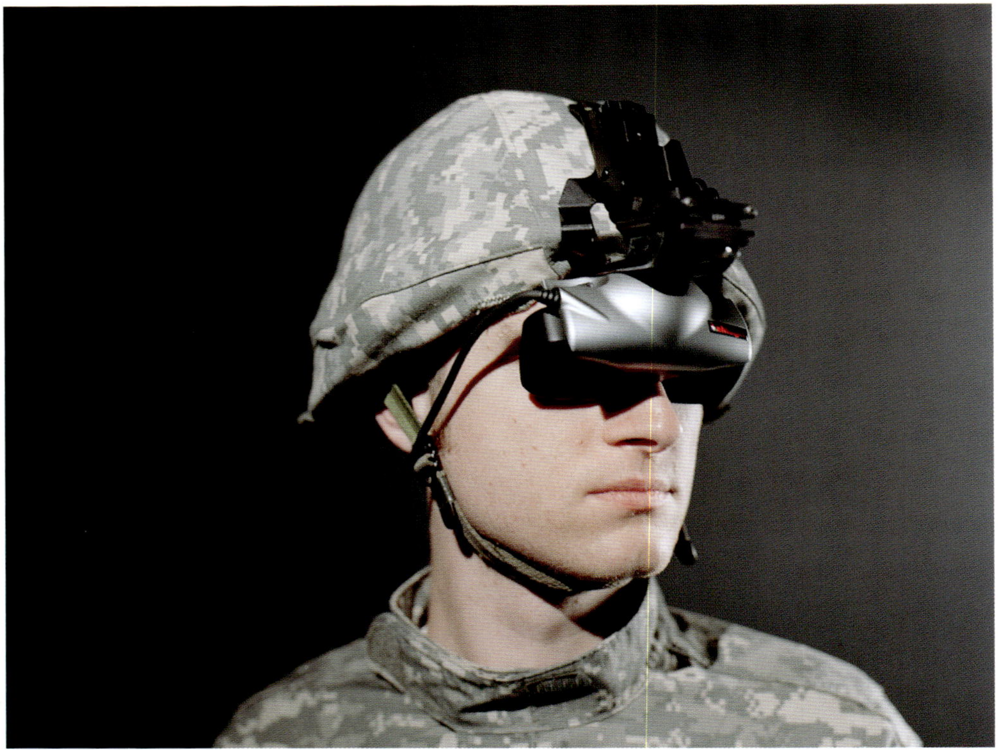

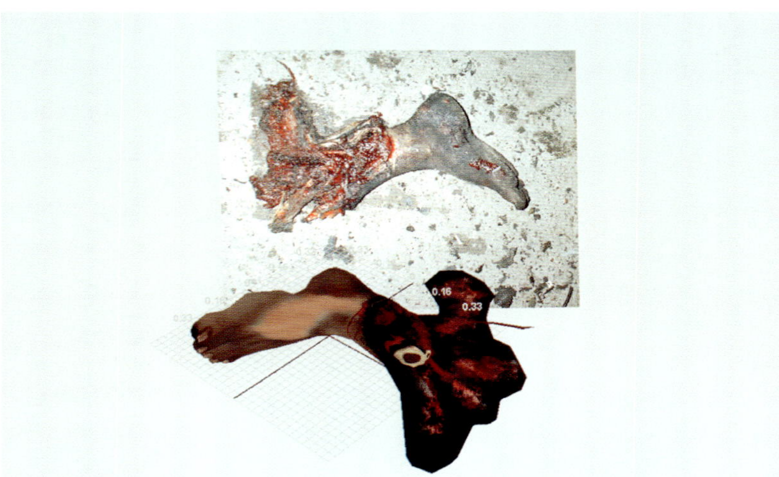

47 Lisa Barnard, *Head Gear*, from the series *Virtual Iraq*, 2008, C-type print.

48 Lisa Barnard, *Virtual Severed Foot*, from the series *Virtual Iraq*, 2008, C-type print.

to make three negatives, which could then be projected, developed and converted by different salts (chromate, nickel and iron) to make yellow, magenta and blue. The result was a very stable colour process that was resistant to fading. At the outset of war, Hennell and his department were seconded to the British Medical Research Council, where they abandoned biscuit tins and instead photographed everything from colour-coded German fuses to the effects of mustard gas on rabbits' eyes.

Hennell's most extensive work was made in collaboration with leading plastic surgeons Harold Gillies and Archibald McIndoe, photographing injured aircrew throughout their long courses of treatment. Hennell's three-quarter profile of a young man with an injury to his nose and left eye (illus. 46) has the surprising dignity of a studio portrait, rather than the forensic character of medical photography. One eye stares steadfastly into the distance; the other is missing, a skin-flap from the bandaged forehead used to cover the wound. Hennell's signature colour process was perfectly suited to recording the fleshy mauves of facial bruising and post-operative recovery. The characteristic softness of the grain lends an unexpected romanticism, even beauty, at odds with the cruel realignment of facial features caused by mechanized warfare. A subsequent image by Hennell shows the patient healing well, his hair shined and parted, dressed in a civilian shirt and tie. It is not known if this young man eventually had his eye artificially restored, but here too Hennell may well have been instrumental: as part of his war work, he photographed 250 varieties of eye colour to assist in the manufacture of glass eyes.

Hennell's images of experimental plastic surgery were produced and used in an instructive capacity, featuring in medical books such as *Atlas of Air Raid Injuries* (1944) and *The Manual of Chemical Warfare*, and remain available for consultation at the British Association of Plastic and Reconstructive Surgeons in London. Recently, however, his images have reached new audiences beyond the medical profession through projects such as the exhibition *How We Are: Photographing Britain*, curated by Val Williams and Susan Bright for Tate Modern in 2007. Repositioned as part of a national history, Hennell's images reveal the unpalatable consequences of war, at odds with conventional patriotic narratives of heroism and soldiering. The episode demonstrates how images originally made in one context may subsequently migrate to unexpected settings and evoke new interpretations: a transformation shared by numerous images in this book.

Experimental photographic technologies continue to support the treatment and rehabilitation of soldiers. The UK artist Lisa Barnard has made an extensive examination of the programmes at the Institute for Creative Technologies, University of Southern California.[80] Her series *Virtual Iraq* (2008) documents methods of virtual reality exposure, used to help prepare soldiers for hostile environments or as therapy for veterans suffering from PTSD (post-traumatic stress disorder).[81] A typical scenario is designed to help helicopter pilots prepare emotionally for arrival at a location where an IED (improvised explosive device) has been detonated. Participants wear a head-mounted display, attached to their helmet, enabling a 360-degree field of vision, supplemented by noises, smells and vibrations that can be safely introduced into the environment (illus. 47). Much of the graphic imagery comes from online sources, or from photographs directly supplied by soldiers on deployment (illus. 48).[82] Barnard's companion series *Triggers* (2008) depicts innocuous urban scenes such as a street with an abandoned shopping trolley, or a pile of rubbish on the pavement, that have become triggers for the PTSD sufferer. For some soldiers returning from war, the battlefield is not left behind, but a psychic state carried with them, even into the backstreets of a familiar West Coast city. Barnard's work eloquently alludes to the fuzzy boundaries between home and front, and the ongoing presence of military activity in civilian spaces. These themes are developed in the next chapter.

three

Home Fronts

Where does war take place? Conventionally, there is a clear division between the realms of military 'front' and civilian 'home'. In practice, however, these distinctions are artificial, and – as the photographs and practices discussed in this chapter show – home and front are inextricably entangled.

The term 'home front' became commonplace in the First World War. The conflict was a 'total war', where entire populaces were mobilized to support their nation's military operations abroad. Home front initiatives, then and since, have ranged from practical efforts such as the fabrication of armaments to the management of more intangible resources such as patriotism and morale. Photography has been persistently deployed in the service of these endeavours, supporting military recruitment, facilitating home-and-front communications, and disseminating propaganda in support of war. Conversely, it has also been an effective tool for civilian activists and dissenters to protest militarism and the wartime actions of governments.

Despite the importance of home-front activities, histories of war – and war photography – tend to prioritize events in the combat zone over the experiences of civilians. Yet civilians have inadvertently found themselves in the field of battle, particularly when revolution, civil war or terrorism erupts in the homeland, and vulnerable to the extended reach of modern weaponry such as long-range aircraft used to decimate civilian targets. Although women, children and non-combatants may face life-threatening dangers of bombardment, homelessness or starvation, they are not generally considered to experience the 'action' of war in the same way as soldiers in the battlefield. Consequently, civilians are often held to have a distinct, and perhaps less authentic, experience of war. Attention to photographs may address this interpretative shortcoming. Images made by, for and about civilians have the potential to enrich our understanding of war as an endeavour that may affect – and be affected by – anyone.

War Effort

The exodus of men to the front in the First World War meant that those ineligible for combat – the elderly, unfit and women – were required to fill positions in essential industries. Mechanized warfare, employed on an unprecedented scale, demanded the mass production of arms. The 'Shell Crisis' of 1915 prompted the British government to make concerted efforts to recruit women, and by the end of the war as many as a million 'munitionettes' had been employed in the production of armaments. In the summer of 1917, the Department of Information appointed Horace Nicholls as Home Front Official Photographer.[1] Nicholls was initially commissioned to provide images for morale-boosting photographically illustrated booklets, and some of his most striking images show women's work in munitions factories, such as Chilwell in Nottinghamshire (illus. 49). Nicholls depicts an impressive regiment of shells, stretching as far as the eye can see, presided over by a series of young women in overalls who perch above their deadly cargo, seemingly unfazed by the dangers of their task. An isolated male worker in the foreground gazes up at them, though whether in curiosity, admiration or trepidation is not revealed. Munitions work was a volatile undertaking: workers lost hands when they accidentally detonated shells, and in 1918 a major explosion at the Chilwell factory left 134 dead. Toxic contamination from chemicals was unavoidable, and women became known as 'Canaries', their skin stained yellow from TNT substance – an unpleasant reality that Nicholls' black-and-white images did not reflect.[2]

Nicholls' photographs became important records of women's participation in the war effort.[3] His images were utilized by the Women's Work Sub-Committee (WWSC), set up in April 1917 under the auspices of the nascent Imperial War Museum. Among other projects, the WWSC commissioned the portrait photographer Olive Edis to document women in the auxiliary services at the Western Front, and produced an exhibition

49 Horace Nicholls, Chilwell munitions factory, c. 1917, silver gelatin print.

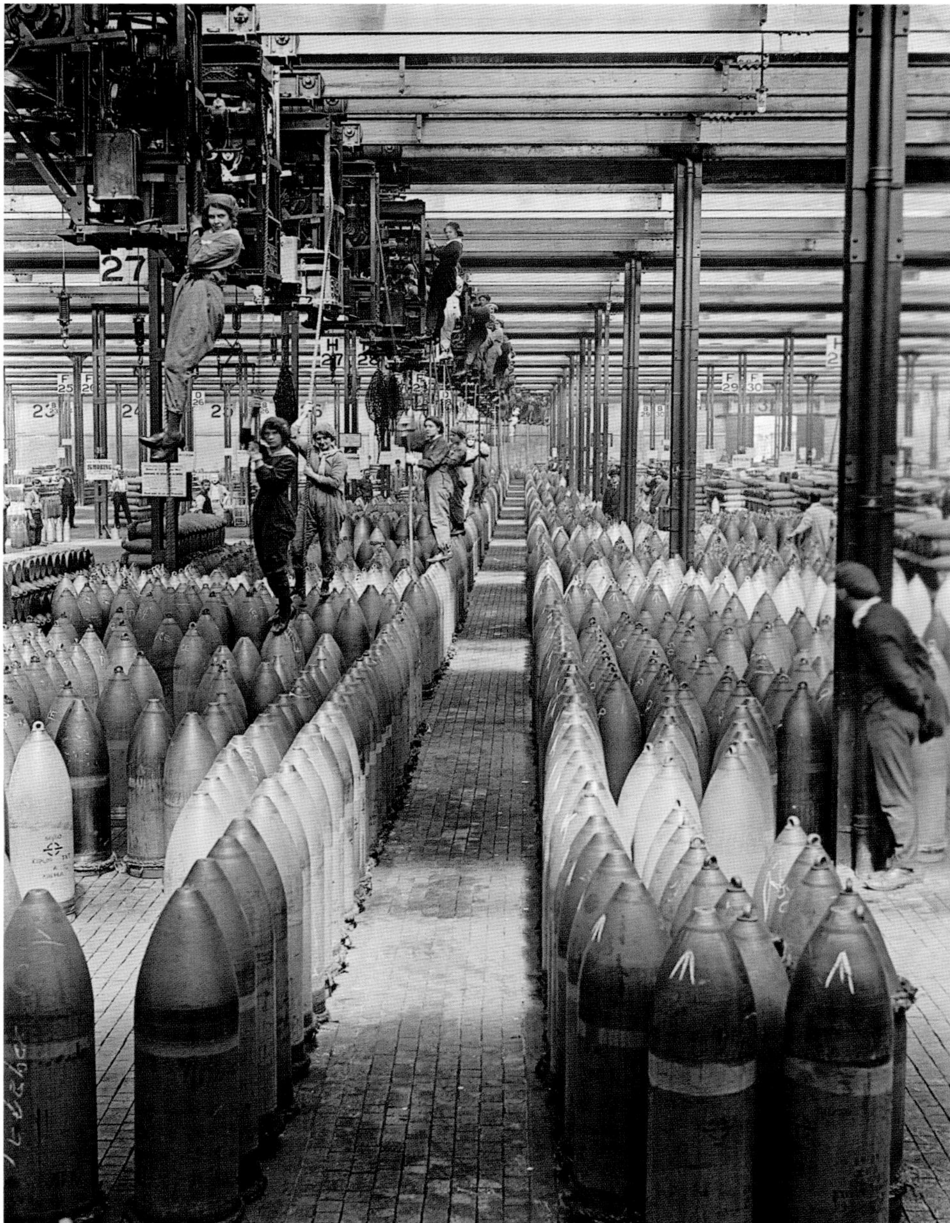
27

entitled 'Women's War Work' at London's Whitechapel Art Gallery in November 1918. Of particular note was the 'Shrine' to women who had died as a consequence of their work during the war. The show received over 82,000 visitors, representing a significant success for the WWSC's mission to spotlight the contributions of women.[4] Besides conveying patriotic support of the national war effort, these photographic projects should be seen against the backdrop of demands for female suffrage, advocating how women had 'earned' their right to full political citizenship.

Gender relations become particularly fraught in times of conflict. It is a paradox that while the rhetoric of war reinforces gender stereotypes – the heroic masculine solider fighting to protect his weaker womenfolk – the social upheaval of wartime has frequently prompted expansions in women's roles. As Nicholls's image of munitionettes demonstrates, the transgression of women in men's attire or undertaking masculine work makes an arresting, and perhaps unsettling, subject. In Germany, the amateur photographer Käthe Buchler made an extensive record of civilian experiences, including the series *Frauen in Männerberufen* (Women in Men's Jobs).[5] This striking series of portraits depicts female bus conductors, coal-haulers and railway workers. One image shows two unidentified women dressed in railway livery (illus. 50). The women's poses mimic masculine postures: one is swamped in a greatcoat and heavy boots, while her companion sports – shockingly – trousers and gaiters.

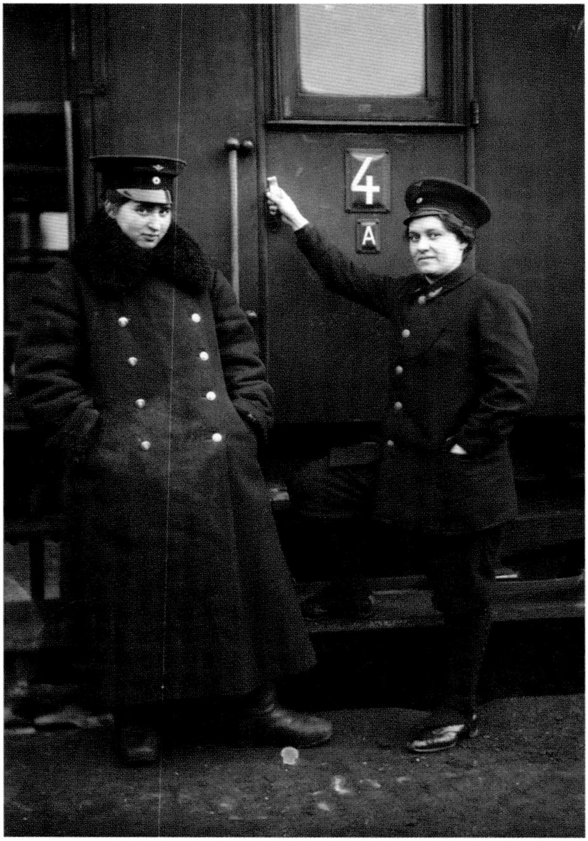

50 Käthe Buchler, train conductors, from the series *Frauen in Männerberufen* (Women in Men's Jobs), c. 1914–18, silver gelatin print.

Elsewhere, photographers have recorded the willingness of women to take up arms in defence of their homes. The Mexican studio photographer Sara Castrejón chronicled the Mexican Revolution (1910–20) when troops occupied her hometown of Teloloapán in the state of Guerrero, an event she initially documented from the safety of an upstairs window.[6] Soon, she adapted to her new clientele.

51 Sara Castrejón, Colonel Amparo Salgado, Teloloapán, Guerrero, Mexico, 1911, silver gelatin print

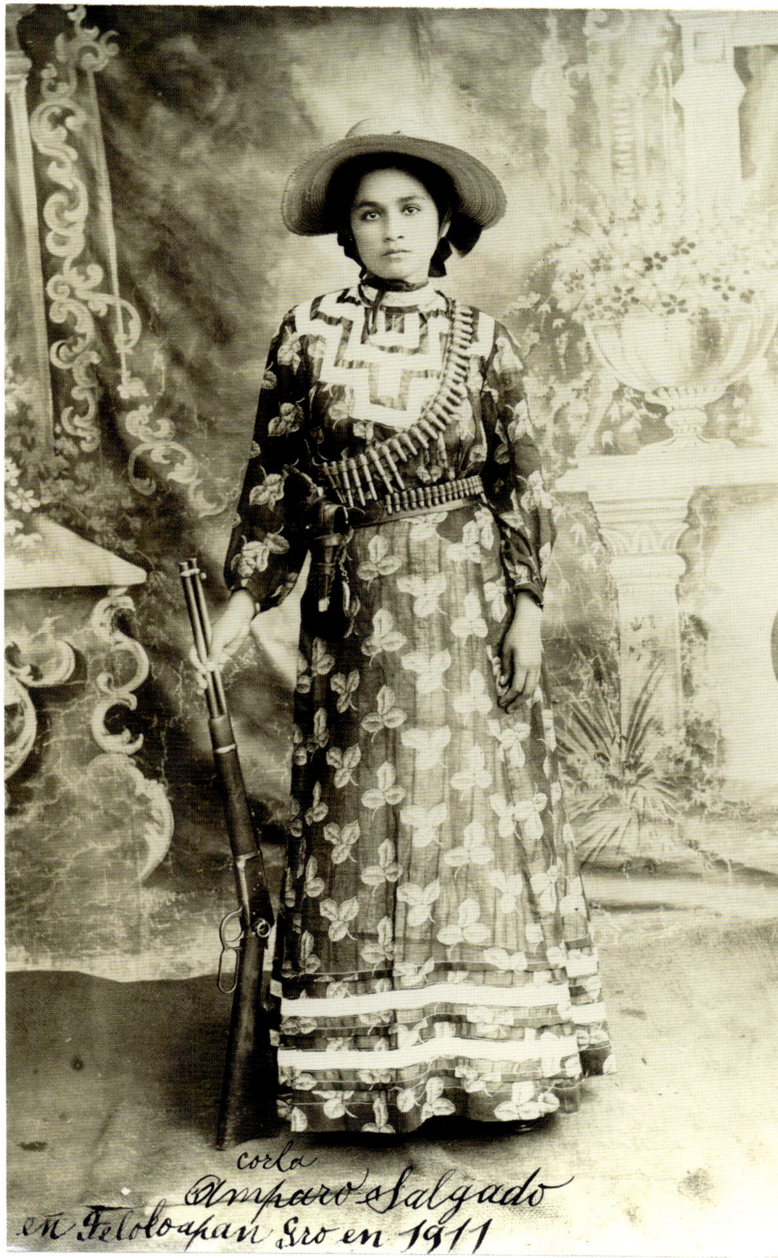

A portrait of the Maderista Colonel Amparo Salgado offers an unexpected collision of visual codes of domestic and military, masculine and feminine (illus. 51). Standing before a genteel studio backdrop, the determined young woman is clad in a ladylike full-length dress that contrasts with her bandolier and rifle. Castrejón's portrait is by no means unique: portraits of armed *soldaderas* (women participants in the Revolution) circulated in the illustrated press as well as on photographic postcards, both in Mexico and the U.S.[7] Although the authenticity of some of these images is questionable, and others are largely symbolic of preparedness to fight, Salgado and a small number of women did engage in combat alongside or within military units.[8] Castrejón also provided photographic services for the occupying forces, recording encampments, manoeuvres and even executions by firing squad. Her work demonstrates not only how 'the front' may be brought unexpectedly to one's doorstep, but the ways in which civilian photographers may become unanticipated chroniclers of war.

Propaganda, Morale and Marketing

Propaganda – the practice of disseminating one-sided information to garner support for one's cause – is probably as old as war itself, but was formalized in the First World War. Most of the belligerent nations set up state or military propaganda agencies, and a progression of departments, bureaux and ministries dispensed material for the Central Powers in Germany, Italy and the Austro-Hungarian Empire, and for the Allies in France, the U.S., the UK and British Commonwealth.[9] A vast and complex practice, which can only be touched upon in this book, war propaganda may be overt or clandestine, based on facts or entirely invented; its target audiences might be fighting soldiers or home-front civilians; allies, enemies or neutrals. Although the term 'propaganda' has acquired negative overtones suggestive of state control, it may be promulgated by any group with a need to persuade, including corporations, the media and protesters. It goes without saying that propaganda functions hand-in-hand with censorship and the suppression of dissenting views, as discussed in Chapter One. While propaganda may be a powerful tool, it must be acknowledged that its efficacy to influence its intended audience is by no means guaranteed (and is often impossible to evidence). Nonetheless, photography has been repeatedly deployed in attempts to stimulate appetite for war, bolster morale or accustom a populace to occupation.

Photographically illustrated magazines have been crucial vehicles for such endeavours, and a handful of examples from the Second World War demonstrate a range of approaches and intended recipients of propaganda. The U.S. photographer Margaret Bourke-White, who had made her name via the hugely popular and influential illustrated magazine *Life*, produced a memorable image for the left-wing, anti-fascist periodical *PM*. Bourke-White could be relied upon to produce dramatic and formally arresting images, and *Flag-making, Brooklyn* (1940) is no exception (illus. 52). Depicting women sewing flags in a factory at the Brooklyn Navy Yard, the formal symmetry of Bourke-White's photograph recalls Modernist photography of the 1920s and '30s. Tightly cropped, the image suggests an infinite number of women, cotton spools and starry folds of fabric, repeated row after row as though in a hall of mirrors. The image is brimming with patriotic fervour: an endless supply of Star-Spangled Banners produced by an army of dutiful American women. Published in the tense period before the U.S. entered the war, the full-page image was titled 'Will these flags be raised against a foreign power?', reflecting the magazine's pro-intervention stance.[10] The observant viewer might notice that one of the operatives is not engrossed in her task, instead surveying the camera unsmilingly: a blip in the otherwise seamless surface of capitalist productivity and obedient female behaviour.

In Japan the government's Cabinet Information Bureau produced an official weekly magazine, *Shashin shūhō* (Photographic Weekly Report), which was sold at newsstands and by subscription. By 1943 half a million copies were produced a week, estimated to reach up to 5 million people. In keeping with the avowed divinity of Emperor Hirohito, kamikaze pilots were presented as warrior-gods, and photographs of these young men, usually pictured from a low camera angle to emphasize their nobility, were regularly featured on the magazine's cover towards the end of the war. As David C. Earhart has shown, civilians, even children, were also urged to resist the enemy and prepare themselves for an honorable death.[11] The 'Kamikaze-fication of the home front' is succinctly expressed in a *Shashin shūhō* cover from 3 January 1945 (illus. 53).[12] The caption states,

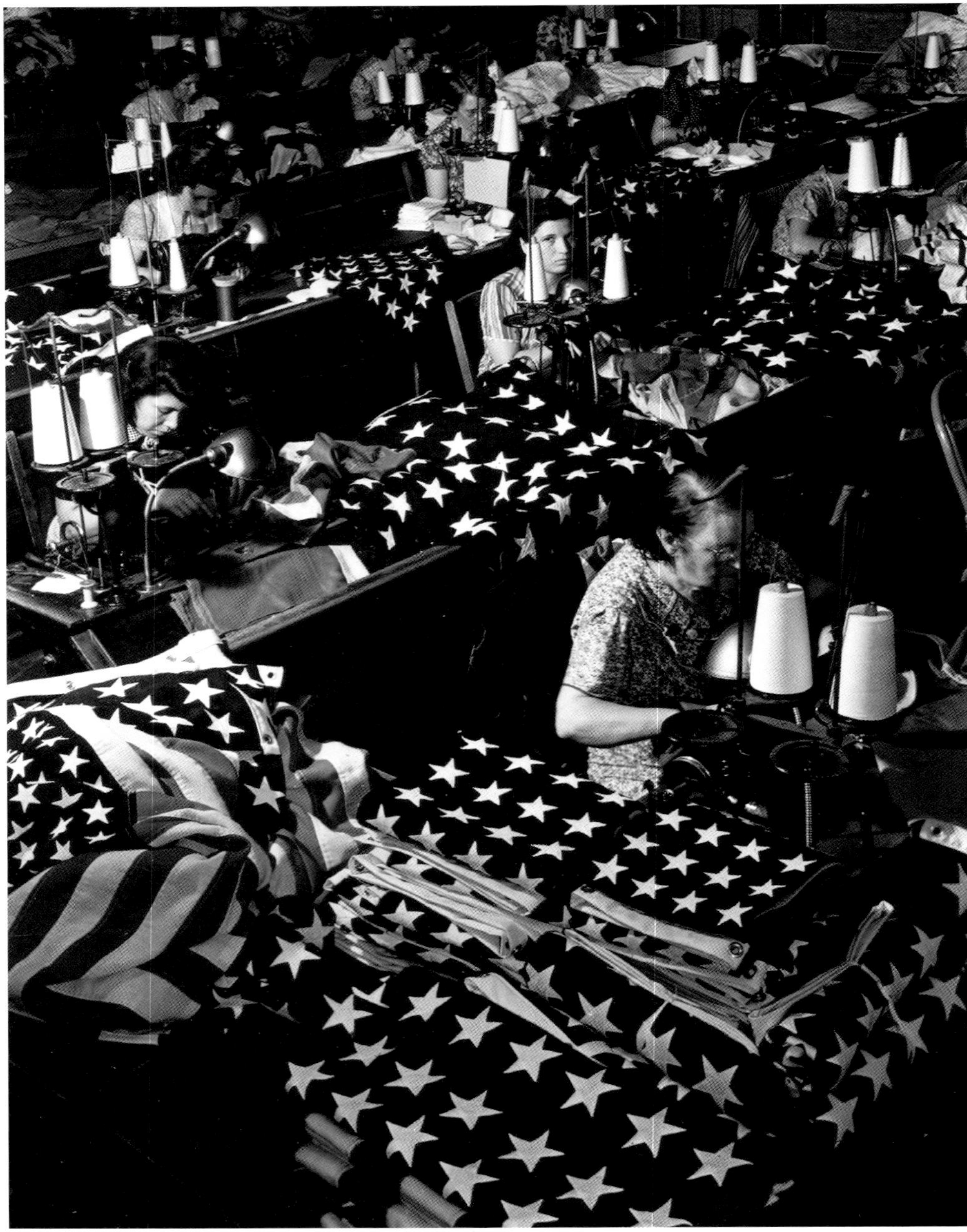

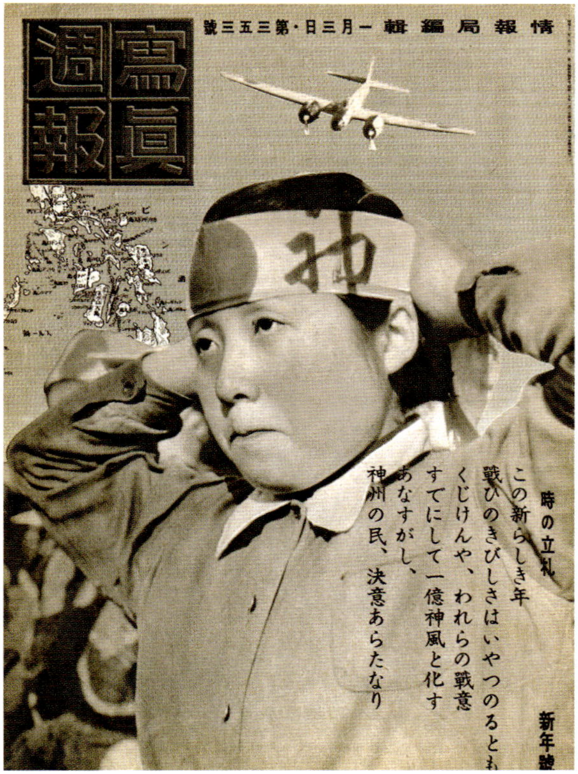

53 Cover of the magazine *Shashin shūhō* (Photographic Weekly Report), 3 January 1945.

> In this new year, although the fierceness of the fight intensifies, our will to fight is unyielding. The one hundred million are already kamikazefied. The people of this Land of the Gods renew their will to fight.[13]

A young girl, her face fixed in grim determination, fixes a headband emblazoned with the word 'Kamikaze'. The image is superimposed with a plane and a map of the Philippines, explicitly presenting the civilian girl's actions alongside those of suicide pilots abroad.

The most widely disseminated state-sanctioned photo-magazine was *Signal*, published fortnightly by the Division for Defence Propaganda under the auspices of the Nazi Wehrmacht from 1940 to 1945. At its height, some 2.5 million copies were circulated per issue, in at least twenty languages, to occupied and neutral countries.[14] *Signal*'s primary objective was to win over hearts and minds: to instil trust in the German forces and to pave the way for a Nazi-led Europe. For most of the war, *Signal* was a model of appealing persuasion rather than political hectoring, and featured articles on art and culture alongside positive news stories of efficient war production and victorious battles. A striking cover from a 1943 Norwegian edition features a handsome Aryan soldier, his clear eyes raised upwards to a noble future (illus. 54). The caption informs the viewer that the soldier is serving in the Russian Defence Corps in Serbia, fighting in the Balkans against 'Bolshevik gangs'. The back cover of a 1944 Danish edition pictures a colourful scene of impoverished Italian children crowding round a German market wagon, from which a fair-skinned serviceman dispenses chocolate. The high production values of the magazine, and particularly its liberal use of rich colour images – a rarity at the time, enabled by the pioneering German film company Agfa – conveyed a confident and technologically advanced nation.

52 Margaret Bourke-White, *Flag-making, Brooklyn*, 1940, silver gelatin print.

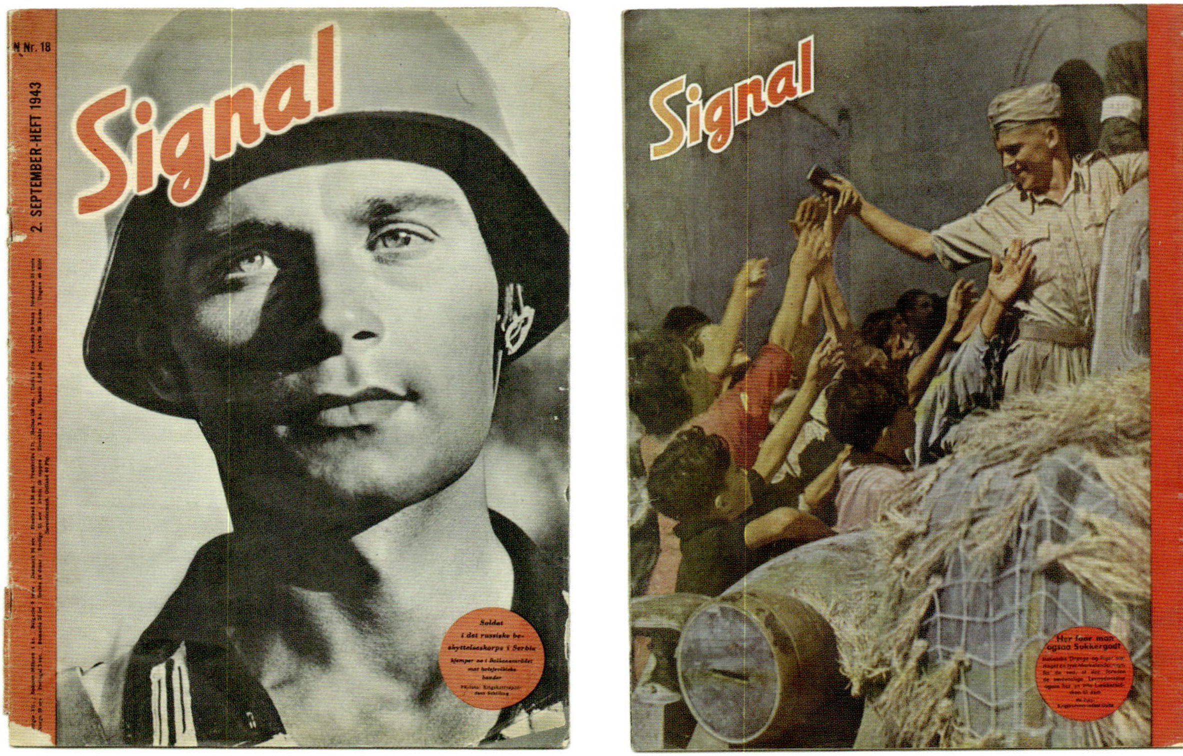

The threat posed by *Signal*'s influence was discussed in the U.S. Congress, and a subsequent report in *Life* magazine (on which *Signal* was partly modelled) claimed that *Signal* was the 'deadliest weapon in the vast Axis propaganda arsenal'.[15] This anxiety may have been unfounded: while *Signal*'s intention to influence was obvious, its reception is much less clear. In France, for instance, vendors in Nice and Lyons refused to stock the magazine, prompting the German embassy in Vichy to order kiosk owners to prominently display *Signal* or have their licences revoked.[16] While the magazine had more success in northern France, that does not mean that *Signal*'s racist and anti-Communist agenda was consumed unquestioningly by its intended audience: many Europeans resorted to the German press simply to keep informed on the progress of the war.[17] Nonetheless, supported by substantial finance and resources, *Signal* trumpeted Teutonic superiority,

54 *Signal* magazine covers. Left: Kriegskorrespondent Schilling, 'Soldier in the Russian Defense Corps in Serbia', Norwegian edn, 18, 1943. Right: Kriegskorrespondent Opitz, 'You get candy here too!', Danish edition, 9, 1944.

55 Doris Davison, photographs made while volunteering for the YMCA Snapshots from Home League, 1942–5, mounted in album. Image amended by author to remove personal details.

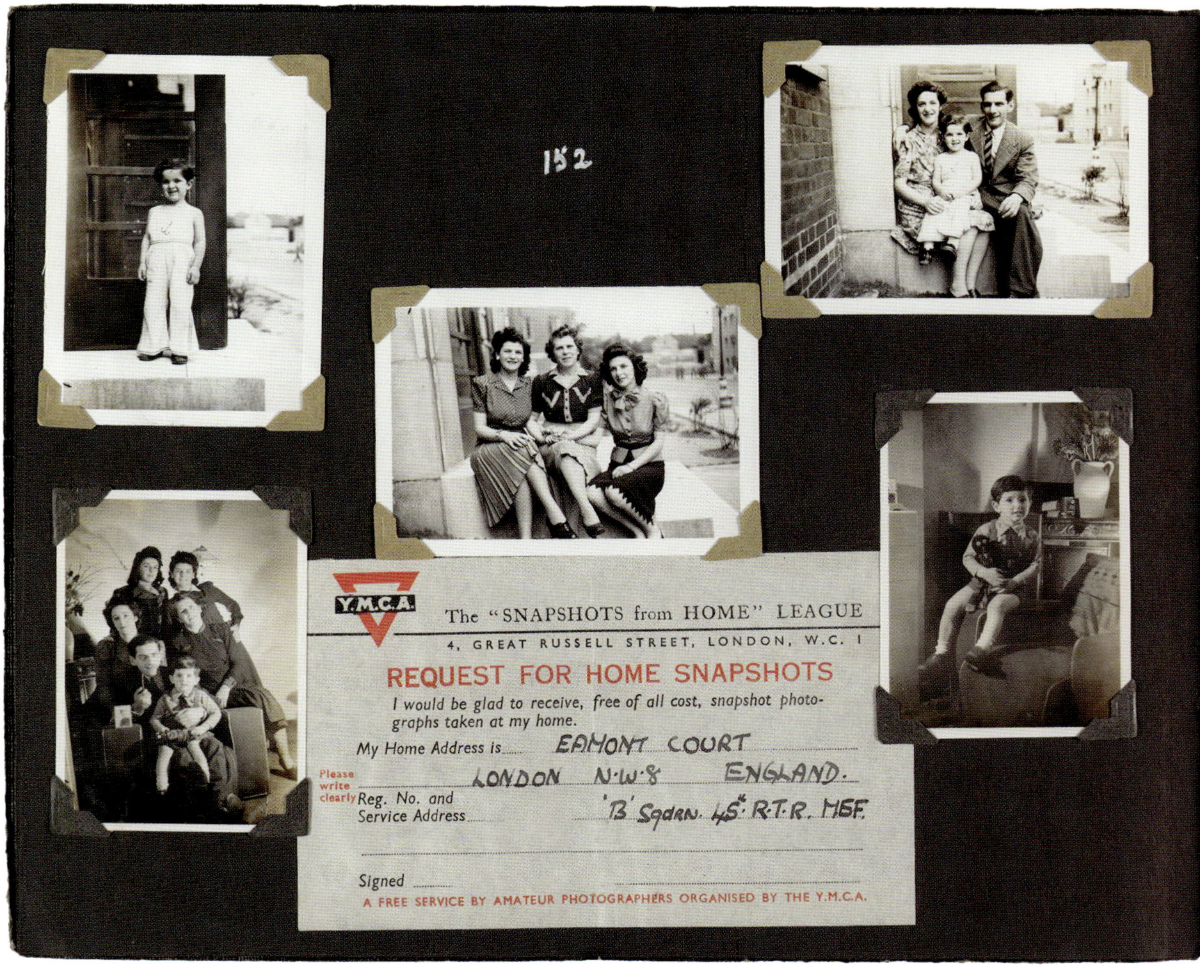

Y.M.C.A.

The "SNAPSHOTS from HOME" LEAGUE

4, GREAT RUSSELL STREET, LONDON, W.C. 1

REQUEST FOR HOME SNAPSHOTS

I would be glad to receive, free of all cost, snapshot photographs taken at my home.

My Home Address is ...... EAMONT COURT

LONDON N.W.8 ENGLAND.

Please write clearly

Reg. No. and Service Address ...... 'B' Sqdrn. 45th R.T.R. MEF.

Signed ......

A FREE SERVICE BY AMATEUR PHOTOGRAPHERS ORGANISED BY THE Y.M.C.A.

promulgating Germanic hegemony in Europe as the best guarantor of peace and safeguard against communism.

As these activities demonstrate, the psychological state of a populace plays a crucial role in warfare. Management of 'morale' – a combination of confidence, enthusiasm and determination – was a priority in both world wars. Personal photography was promoted as a vital way to motivate and comfort servicemen and women stationed away from home. The charity YMCA ran national photography campaigns in the UK and U.S. under the banner 'Snapshots from Home', whereby soldiers and war workers could request photographs of loved ones to be sent.[18] Volunteer civilians, usually older men unfit for service or leisured women, visited spouses and families and sent the resulting snapshots overseas. One such volunteer was Doris Davison, a British woman who worked for the service between June 1943 and November 1945. As a memento of her work, she compiled an album that includes request forms filled in by servicemen, copies of the snapshots sent in return and a number of thank-you letters from the recipients (illus. 55). A valuable record of an overlooked form of photographic participation in war, the album is touching and at times heartbreaking, in the case of packets returned unopened having reached their recipients too late.

Kodak firmly supported the Allied war effort and, in a canny blend of patriotism and commercialism, presented civilian snapshot photography as a righteous endeavour. A magazine advertising campaign in the First World War encouraged women to record their husband or son on 'The Day of His Leaving', or 'Before He Goes'.[19] Autographic cameras, which enabled the user to record the date on which the photograph was taken, were promoted as a means of making an accurate historical record. By the Second World War, Kodak's emphasis had shifted from anticipated memories to here-and-now communication, urging consumers to 'visit your man in the service with snapshots'. Marketing campaigns in popular magazines such as *Life* and *Collier's* claimed that snapshots were a sure-fire way to boost a man's morale and, by implication, his motivation to fight.[20]

One of the most explicit applications of persuasive home-front photography has been its use in recruitment campaigns. When conscription has not been enforced, aspirational images have long been a potent means of attracting would-be soldiers. While posters and advertisements from the

56 Saatchi & Saatchi, recruitment poster commissioned by the British Army Careers Service, c. 1998, colour photolithograph.

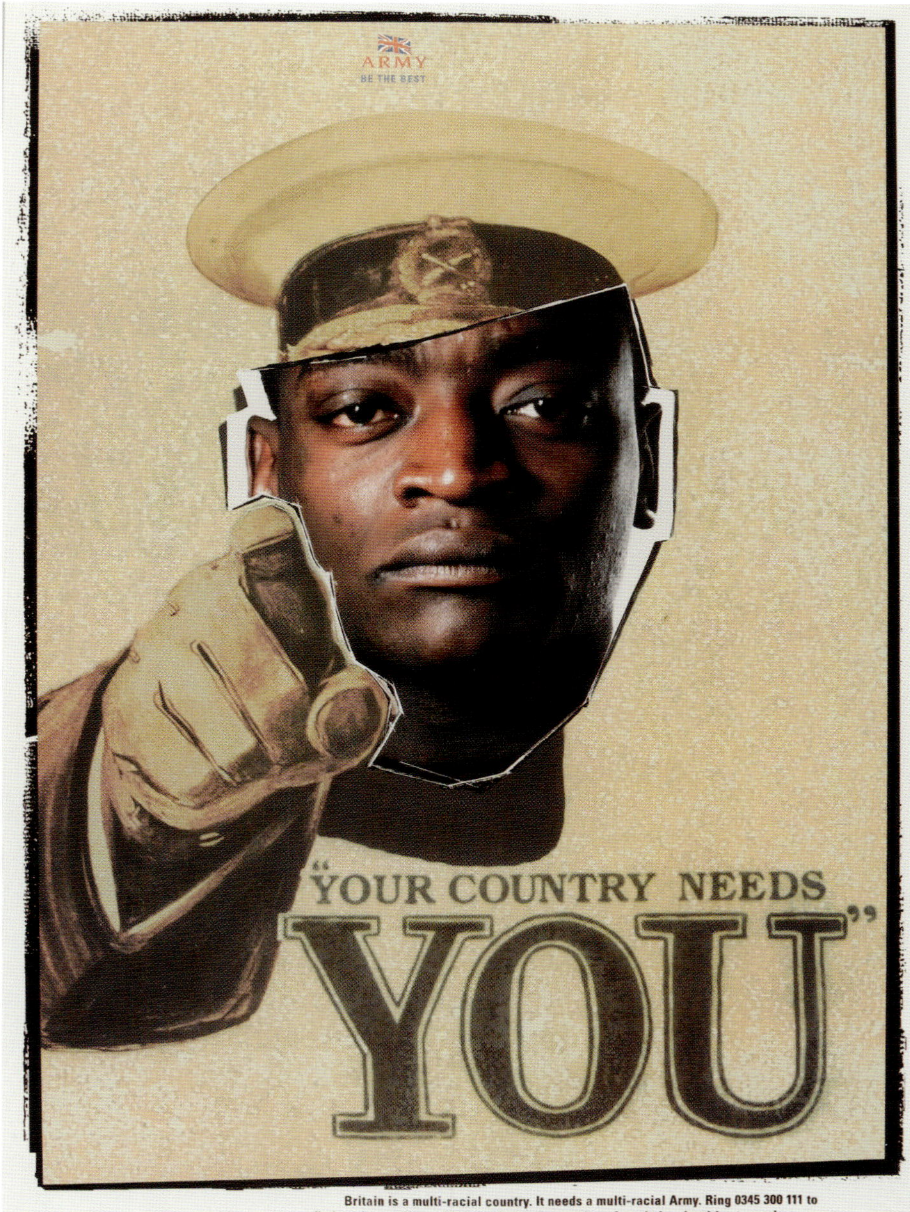
ARMY
BE THE BEST
"YOUR COUNTRY NEEDS
YOU"
Britain is a multi-racial country. It needs a multi-racial Army. Ring 0345 300 111 to
find out how we're making the Army a better place for ethnic minorities to work.
Or call 01252 334 850 to talk directly to a Black or Asian soldier.

nineteenth century and the first half of the twentieth century were likely to feature illustrations, such as the First World War poster depicting the bombing of Scarborough (see illus. 62), photographs increasingly became central to the marketing initiatives of armed forces.

Instigated by the New Labour government in 1998, the British Army Careers Service commissioned the advertising company Saatchi & Saatchi to produce a recruitment campaign that aimed to address both the chronic shortage of soldiers and accusations of racism in the army. A bold poster from the campaign features Alfred Geete's iconic illustration of Lord Kitchener (Secretary of State for War in 1914) replaced by a roughly clipped photograph featuring the face of Ghanaian-born Captain Fedelix Datson of the Royal Artillery (illus. 56). Datson's uncompromising gaze, looking downwards at the viewer in a position of power, connotes strength, dignity and resolve. As sociologist Vron Ware notes, the poster unabashedly aims to stir patriotism and nationalism in those who might otherwise have felt excluded in postcolonial Britain.[21] While high rates of unemployment for working-class men and other sociological factors certainly contributed to enlistment, the advertising campaign garnered awards for helping achieve the Army's target of 15,000 Black and minority ethnic soldiers. Whether institutionalized racism in the armed forces was addressed is another matter entirely.

The apparently inevitable and desirable transformation of boys into soldiers has been critiqued by some photographers, who highlight the ways in which militarism is marketed and, indeed, culturally embedded in society. Melanie Friend, whose work *The Home Front* (2009–12) is discussed later in this book, has drawn attention to civilian and military air displays as sites of entertainment and recruitment. Jill Cole, also in the UK, has made an ongoing study of a military training ground near her home in northeast England. *Paintball Portraits* (2007–8) depicts boys and young men playing action games in the first few weeks of training (illus. 57). Cole's sensitive portraits accentuate fair eyelashes and adolescent downy skin spattered with orange paint, a proxy for blood that these young men may one day shed. A companion series, *Birds* (2007–8), responds to the training ground's status as a protected natural habitat, and pictures common birds such as chaffinches and robins entangled in netting before being logged and ringed. As well as suggesting feelings of entrapment – perhaps alluding to the limited

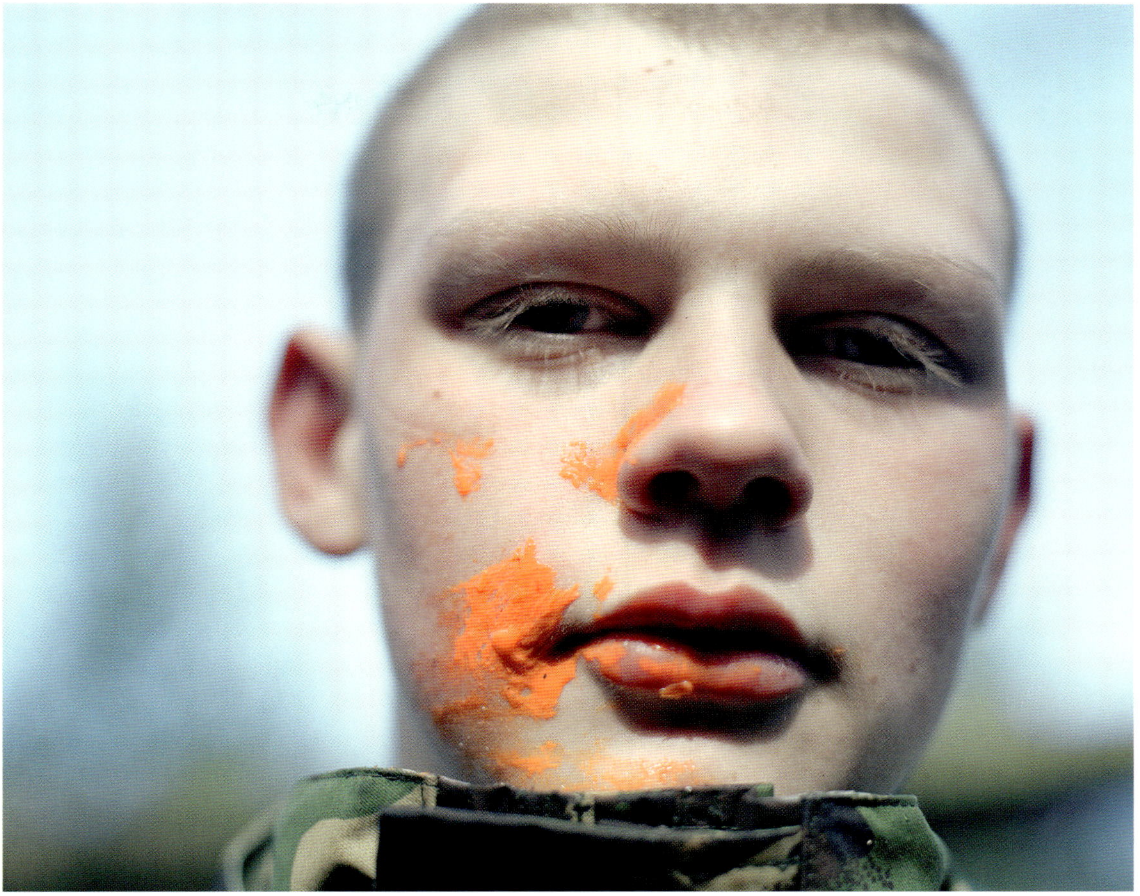

57 Jill Cole, *Paintball Portrait #1*, 2007–8, C-type print.

career options of working-class boys in the area – Cole's series connects with criticisms of 'greenwashing', whereby military occupation is made more palatable by a veneer of environmental concern. As a whole, her photographs aim to investigate the ongoing presence of war and the ways in which home may become militarized by stealth.[22]

Cole's work points to the problem of defining 'home' and 'front' by geographical boundaries. While the homeland is assumed to be a safe place, far from the fighting of troops in distant theatres of war, certain kinds of armed conflict, such as revolution, civil war and terrorism, unfold in the heart of civilian territory. The veteran British war photographer Don McCullin, who produced iconic images of foreign wars including *Shell-shocked U.S. Marine, Vietnam* (1968) and *Albino Orphan, Biafra* (1969), was assigned by the *Sunday Times* to photograph in Northern Ireland in the early 1970s. McCullin made some of the most memorable images of The Troubles, a conflict marked by paramilitary action, guerrilla warfare and sectarian violence. *The Bogside, Derry* (1971) shows soldiers of the Royal Anglian Regiment counter-attacking stone-throwers in the Bogside neighbourhood (illus. 58).[23] This Irish Republican/Catholic-majority zone, bordering an Ulster Loyalist/Protestant enclave, became a notorious hotspot during The Troubles. McCullin's image captures the drama of soldiers equipped with riot shields charging down a street where two women are tucked into doorways. The armed, uniformed men form a strong contrast with the female civilians who, at first glance, might be assumed to be cowering, but the image reveals a more complex interplay. A lopsided helmet undermines the authority of the soldier on the left, while the impassive older woman seems to treat the eruption of violence as an everyday event, and the younger woman appears to be shouting abuse at the British forces. The image was one of several published as 'War on the Home Front' in the *Sunday Times* colour supplement, one of the pre-eminent sites for photojournalism in the 1970s, bringing the conflict vividly before British viewers.[24]

In addition to the contributions of professional photographers, civilians have frequently been important chroniclers of the effects of conflict on ordinary lives, without ever considering themselves bona fide war photographers. It is hard to imagine a more literal or shocking collision between 'home' and 'front' than the tragedy of 9/11, when two of four planes hijacked by Islamic terrorist group al-Qaeda were flown into the Twin Towers of the World Trade Center in Manhattan on 11 September 2001. The scenes of devastation were recorded by prominent professional photographers including

Joel Meyerowitz and James Nachtwey. The proximity of bystanders to the scene, however, coupled with the availability of cameras both analogue and digital, generated a deluge of non-professional photographic imagery by residents, city workers, tourists, volunteers and emergency personnel. For many, taking pictures may have been a way of processing the traumatic event as part of a collective endeavour. One of the most prominent shared sites for photographs was the spontaneous display that sprouted in two storerooms on Prince Street in SoHo, downtown Manhattan (illus. 59). Begun in early October, *Here is New York: A Democracy of Photographs* exhibited around 4,000 images contributed by five hundred photographers, both professional and non-professional. The contributors to *Here is New York* expected no financial return: the proceeds of print sales were donated to various charities. Displayed simply on wires hung across the space, the images were shown without

58 Don McCullin, *The Bogside, Derry, Northern Ireland*, 1971, silver gelatin print.

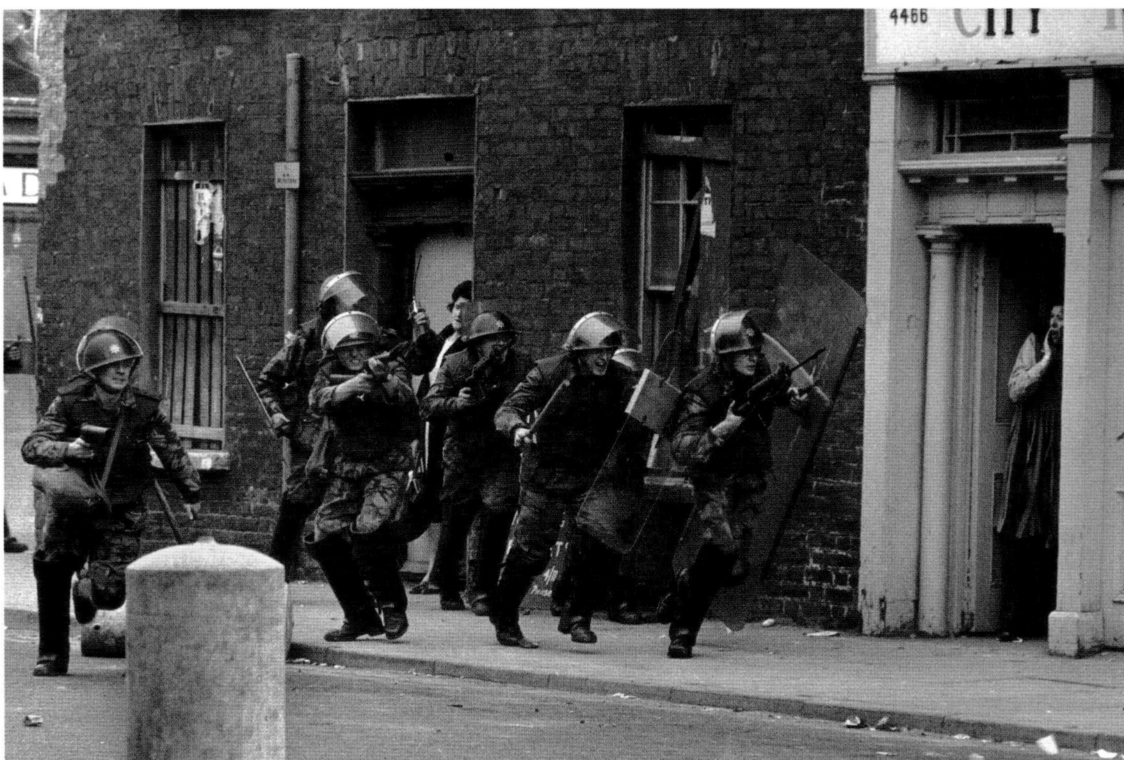

captions or attribution, presenting 'history in a new egalitarian way', according to the *New York Times*.[25] Selections from the display travelled to satellite shows in the U.S. and numerous international cities including London, Berlin and Tokyo, becoming a kind of travelling memorial: a collective whole composed of many individual stories and viewpoints.[26]

Bystander photography of the 'front' rupturing 'home' is almost as old as photography itself. Perhaps the first conflict to be photographed in this way was the violent uprising in Paris in the 1848 Revolution, part of a tumultuous year of unrest in Europe. M. Thibault, a French photographer of whom little is known, photographed rue Saint-Maur-Popincourt from a high vantage point, probably an attic window (illus. 60). Cobblestones have been torn up to barricade the street, which seems to be deserted in the tense interim before or after violence (it was made, in fact, prior to General Lamoricière's repression of the uprising). Thibault's daguerreotypes, replete with quotidian details such as the painted shop sign advertising chocolate, conveys the eerie transformation of home territory into combat zone. On close inspection, a woman wearing a white cap can be seen leaning out from an upper-ground-floor window, peering at the scene through a louvred shutter. Thibault's identity remains obscure, but he appears to have been an innovative amateur

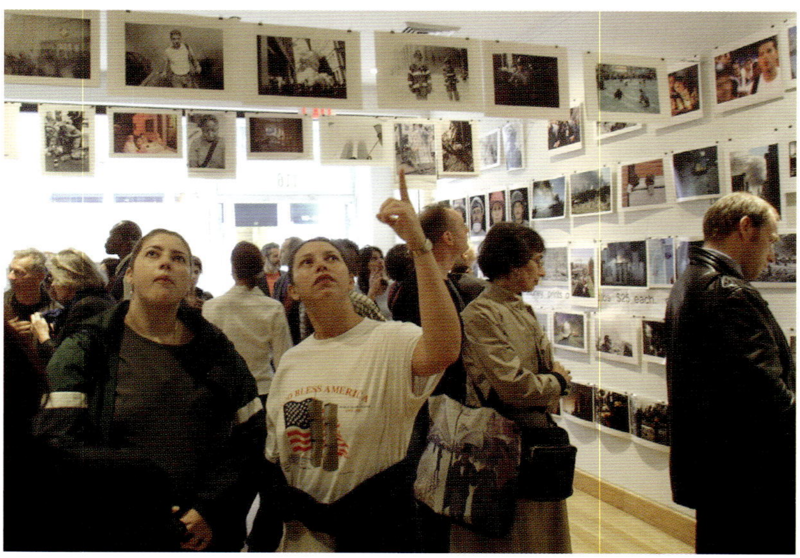

59 Doug Kanter, visitors to the exhibition *Here is New York*, SoHo, New York City, 2001.

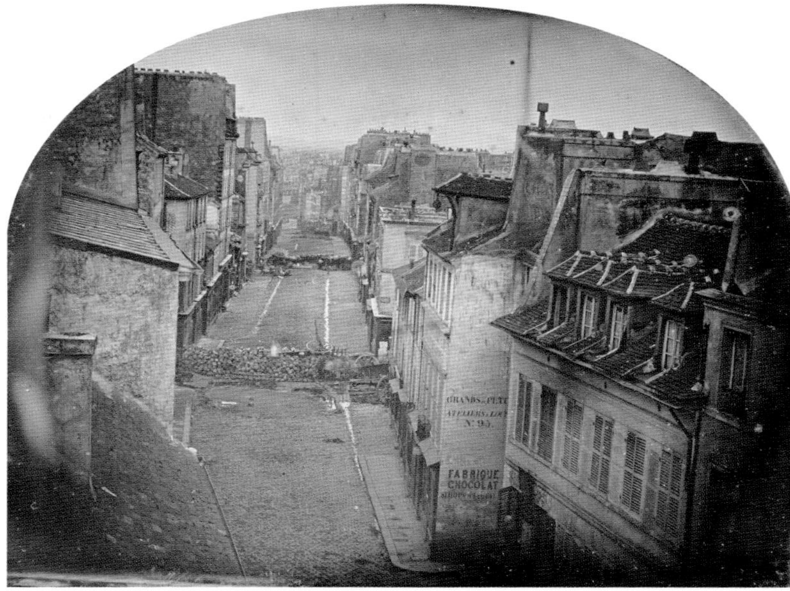

60 M. Thibault, barricade at rue Saint-Maur, Faubourg du Temple, Paris, 1848, daguerreotype.

photographer responding to extraordinary events taking place in the heart of the city, a kind of prototypical citizen journalist. In what is arguably the first instance of photojournalism, two of Thibault's images (depicting the street 'before' and 'after') were used in the French weekly newspaper *L'Illustration: Journal Universal*. Owing to the limitations of print technology, the daguerreotypes had to be translated into wood engravings in order to be reproduced. Captioned 'd'après une planche daguerréotypée par M. Thibault', the illustrations emphasize their photographic origins and their claim to be genuine depictions of events.[27]

By the First World War, long-range weaponry fired from ships, Zeppelins and other aircraft brought civilian targets within reach of distant forces.[28] The German naval bombardment of the east coast of England in the First World War resulted in hundreds of civilian deaths and sparked a national outcry. Fred Foxton, a now-forgotten studio photographer in Scarborough, responded quickly to the devastation of 16 December 1914 by photographing the house at 2 Wykeham Street where Mrs Barnett and her two children were killed. The public was eager to view the spectacular and still novel effects of modern warfare, and Foxton's image appeared ten days later in *The War Illustrated* as

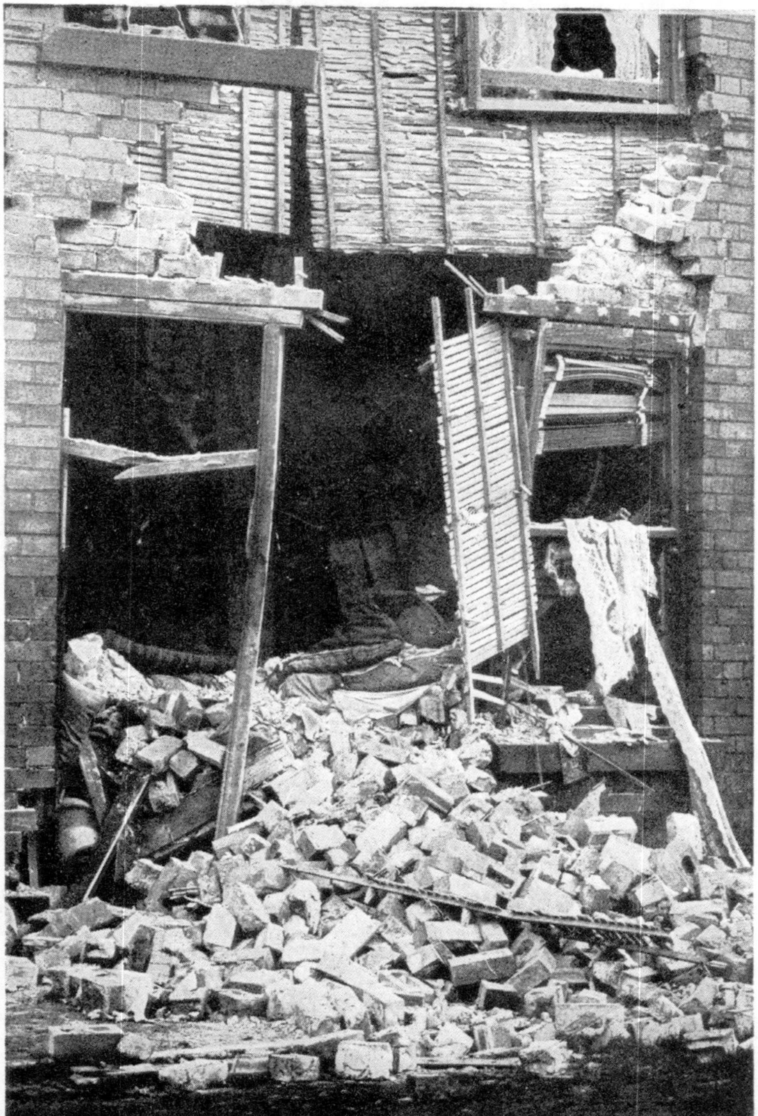

BOMBARDMENT OF SCARBOROUGH.
House in Wykeham Street, where Mrs. Barnett and two children were killed.

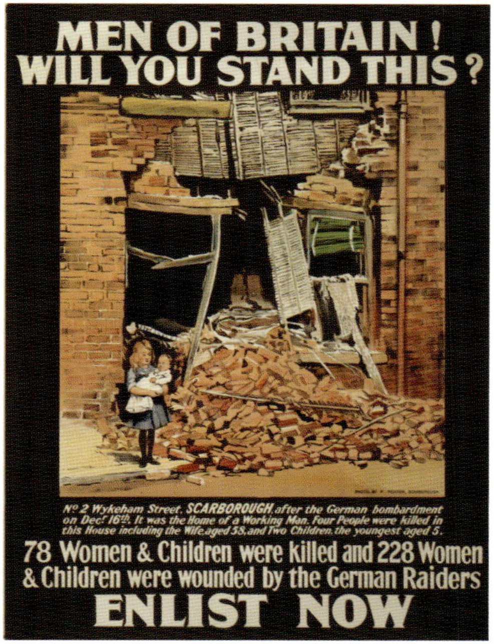

62 'Men of Britain! Will You Stand This?', poster designed and printed by Johnson, Riddle & Co., London, after a photograph by F. Foxton, published by Parliamentary Recruitment Committee, London, 1915.

part of the emotively titled article 'Germans Attack "the Englishman's Home" at Last', and was also circulated as a photographic postcard (illus. 61).[29] The image was later put to use by the Parliamentary Recruitment Committee in their 1915 poster proclaiming 'Men of Britain! Will You Stand This?' The illustration is clearly based on Foxton's photograph, with the blatant addition of a little girl holding a baby in her arms, next to the rubble of the bombed house (illus. 62). The circulation of the Foxton image in its various formats urged viewers to recall the danger to ordinary civilians thousands of kilometres from the enemy, and hence the necessity of military action – and enlisting.

Photographic depiction of the bombardment of civilians became particularly charged as military technology reached new levels of sophistication with aerial weaponry, concurrent with the rise of photojournalism and the photographically illustrated press. In the Spanish Civil War, the spectacle of aerial devastation meted out by Franco's fascist air force, supported by the German Luftwaffe, prompted emotive imagery of terror-struck civilians (not least Picasso's painting *Guernica*). A group of international anti-fascist photographers – among them Robert Capa, Gerda Taro, David 'Chim' Seymour and Margaret Michaelis – collaborated on the publication *Madrid*, published by the Catalan Propaganda Commisariat in 1937. The cover (illus. 63) is a photomontage featuring a now-iconic image of civilians by Chim, closely cropped to focus attention on a mother who seems to anxiously scan the skies even as a baby feeds at her breast. Like many icons of war photography, it should not be taken at face value. As photo-historian Erika Zerwes has observed, the image does not depict an air raid, but in fact shows attendees listening to speakers at a land meeting in Extremadura in 1936, two months before the war began.[30] Divorced from its original context, Chim's image is superimposed on a dramatic background of clouds, to which have been added the silhouettes of a fleet of fighter planes, and a huge bomb adorned with a swastika.[31] Published in four languages (Catalan, Spanish, French and English), the

61 Fred Foxton, *Bombardment of Scarborough*, 1914, photographic postcard.

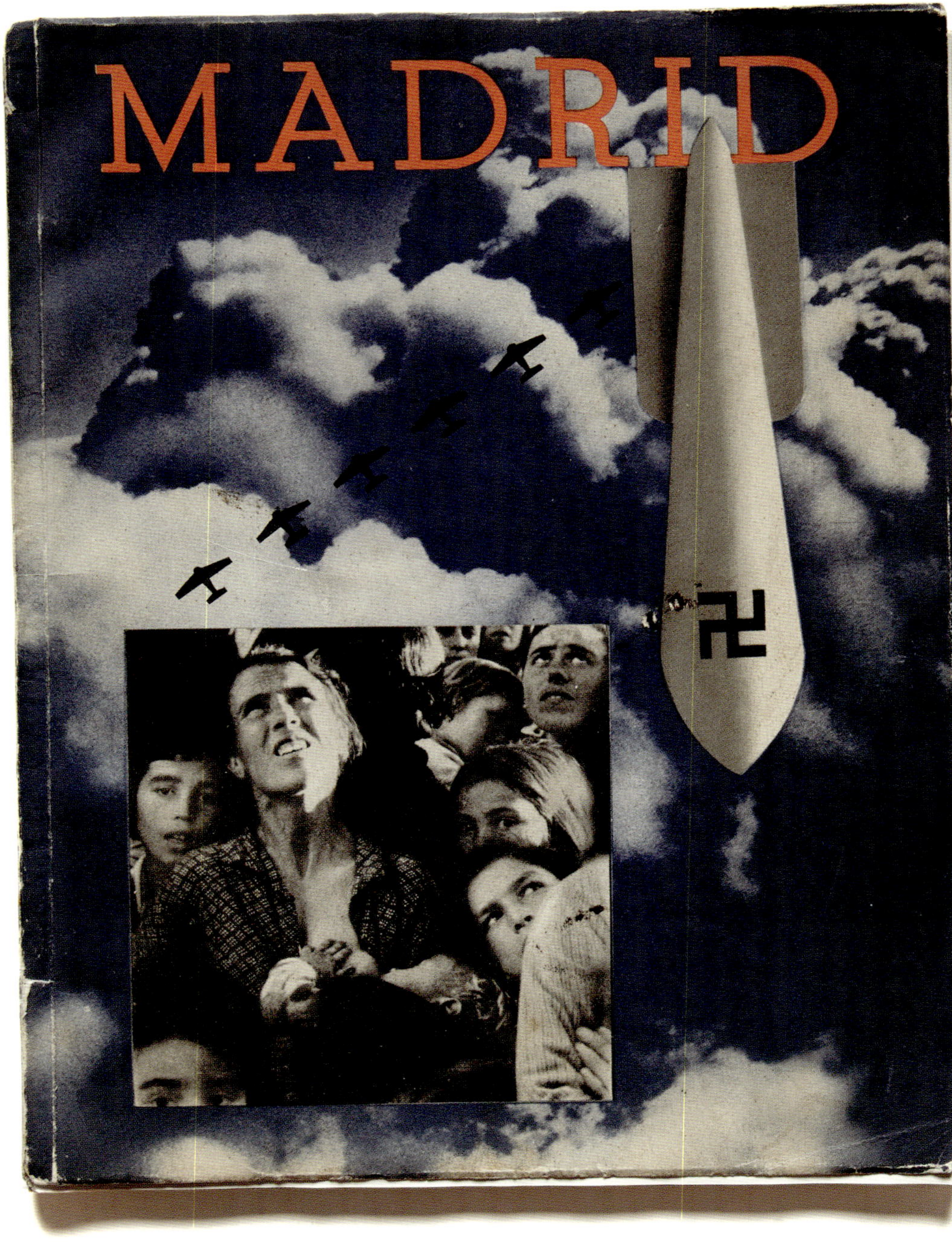
MADRID

book had a clear propaganda role in presenting evidence of Fascist aggression and urging collective international military intervention.

As the cases of the Scarborough poster and *Madrid* book demonstrate, the suffering child is a recurrent motif in the photography of war. Some of the most iconic images are in this mould: Cecil Beaton's portrait of Eileen Dunne, a young girl injured in the Blitz in 1940; H. S. Wong's 1937 picture of an orphaned Chinese baby in a bombed railway station during the Sino-Japanese War; and Nick Ut's 1972 image of a young Vietnamese girl fleeing a napalm attack (see illus. 14), to mention only a handful. Such images appear in the mass media only if it is expedient to do so. It is hard to imagine, for instance, *Life* or *Picture Post* publishing pictures of German children killed by Allied bombs; on the other hand, the timing of *Accidental Napalm*, which depicted an airstrike by the U.S.-aligned South Vietnamese force, coincided with national disenchantment with the Vietnam War. The repeated use of images of suffering children therefore attests to not only the tragic consequences of war on children, but the ongoing currency of the trope as a means of eliciting strong emotional reactions for a variety of purposes.

In the context of humanitarian aid work, the civilian child is particularly effective in evoking universal family bonds, and by extension charity and compassion. The U.S. photojournalist Thérèse Bonney, who had documented the Russo-Finnish War of 1939–40, devoted herself to raising awareness of the plight of children. A press release to announce her exhibition *War Comes to the People*, held at the Museum of Modern Art in New York in 1940, declared that in modern warfare 'no rules or precedents hold and the home fires become the battle front.' Supported by the Carnegie Corporation, Bonney's intention to return to Europe to photograph the effects on civilians was clear:

> She is not going as just another war photographer. She is not interested in what the war does to or with soldiers. She will focus her lens on the aftermath of war – its effect on people, their lives, their homes, their possessions, their hope and their despair.[32]

The exhibition was only one of Bonney's efforts. She worked with the Red Cross and other relief organizations and staged exhibitions at the Library of Congress and the Coordinating Council of French Relief Societies, and her

63 Cover of the publication *Madrid* featuring an inset image by Chim (David Seymour), 1937.

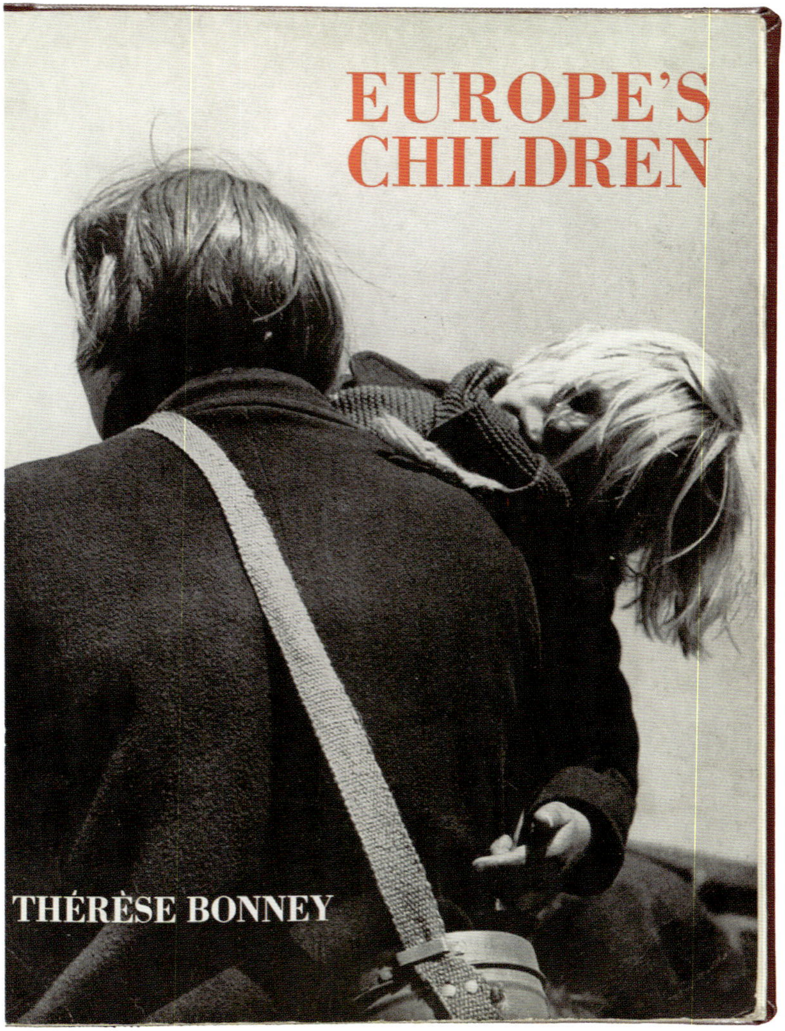

64 Thérèse Bonney, cover of the publication *Europe's Children*, 1943.

photographs appeared in *La Revue de la France Libre* and U.S. *Vogue*. An exhibition toured forty U.S. cities, and her self-published photobook, *Europe's Children* (1943), sold out almost immediately (illus. 64).[33]

Bonney's images of exhausted children fleeing danger were intended to inspire sympathy in countries that might be resistant to an influx of war refugees. Sadly, this plight continues. One of the most affecting of recent

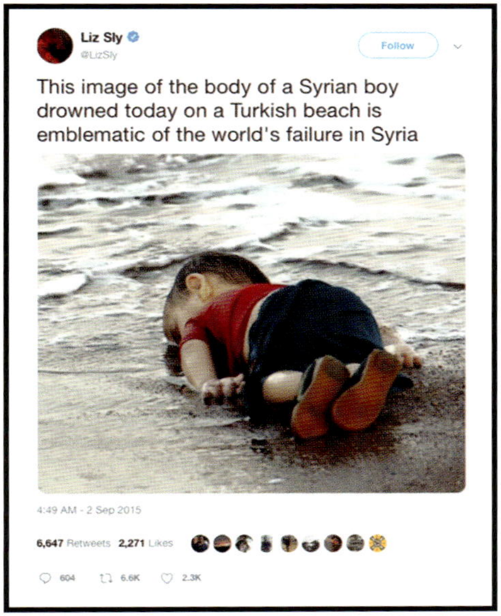

65 Tweet by Liz Sly featuring Nilüfer Demir's photograph of Alan Kurdi, Bodrum, Turkey, 2 September 2015.

cases is that of Alan Kurdi, the drowned three-year-old Syrian refugee, whose body was washed up on a tourist beach at Bodrum, Turkey, on 2 September 2015. A series of images taken by Turkish photojournalist Nilüfer Demir for the Dogan News Agency were circulated via the social media network Twitter, initially by Turkish, Middle Eastern and Arabic users, before going viral and reaching global viewers. One of the first tweets to gain traction came from Liz Sly, Beirut Bureau Chief for the *Washington Post* (illus. 65).[34] The image she chose to tweet – a close-up of the small body of the child, without any aid workers present – emphasized his abandonment, described by Sly as 'emblematic of the world's failure in Syria'. While the authenticity of the child's status as a war refugee was disputed by some, the potency of the image in eliciting mass sympathy and practical aid was undeniable. The images were published on the front pages of newspapers across the world as well as sparking a whole series of memes via social media, performance art, protest and demonstrations, satirical cartoons and photomontages.

The fact that Syrians were dying in their attempts to escape intolerable conditions was not in itself a revelation, so why did the case of Alan Kurdi stand out? Perhaps because the images were not visceral – the little boy could be mistaken for merely being asleep – they were deemed acceptable for publication and sharing. Perhaps also, as humanitarian scholars Fehrenbach and Rodagno have asserted, Kurdi's pale skin and neat appearance, his shoes still carefully fastened, suggested a 'deserving' migrant that might elicit empathy from white Western audiences.[35] Certainly the right-wing media in the UK made an abrupt U-turn in their demonization of refugees, switching terminology from 'cockroach migrants' to 'tragic tot'.[36] The pathos of the child's lifeless body prompted commentary from the world's political leaders and a huge increase in public discussion of the refugee crisis on social media, as well as practical responses. The Malta-based charity Migrant Offshore Aid Station, for instance, reported a fifteenfold increase in donations, while the Charities Aid Foundation in the UK reported that a third of donations to refugee aid organizations had been sparked by Demir's images.[37] While the long-term effects on policy have yet to be seen, for many viewers, the photographs spoke to them as an individual human tragedy to which they could relate more effectively than group statistics.

Civilian Protest

Not all civilians are content to support war and its systems. Activists and politically minded artists have repeatedly made use of photography to oppose militarism or criticize their governments' actions. The German pacifist Ernst Friedrich, whose post-First World War project *Krieg dem Krieg!* (War Against War!) is discussed in the next chapter, published and exhibited graphic photographs that might dissuade citizens from supporting future armed conflicts (see illus. 89). Other German dissidents made effective use of photomontage, the re-assemblage of elements clipped from pre-existing photographs to create hybrid and often startling images. Photomontage was by no means new, but in the early twentieth century the technique became a radical tool to sharply critique society, war and politics, most notably in Weimar Germany in the prelude to the Third Reich. Hannah Höch, Marianne Brandt, John Heartfield and George Grosz disrupted the naturalistic surface of photography with their works. In place of a seamless portrayal of reality, which suggested that social order was natural and unalterable, their photomontages foregrounded the construction of images, suggesting that structures and systems were man-made and open to disruption.

John Heartfield was a committed political activist who had anglicized his name from Helmut Herzfeld in a protest against anti-British feeling in Germany in the First World War. Many of his photomontages of the 1930s appeared on the covers of the communist weekly *AIZ* (*Arbeiter Illustrierte Zeitung*, the Workers' Illustrated Newspaper) and rotogravure posters on the streets of Berlin.[38] His overt and vitriolic attacks on the National Socialists are still astonishing in their power. An *AIZ* cover from 1932 condemns the impotence of international diplomacy to constrain Hitler: the dove of peace is brutally skewered on a bayonet, while in the background the Swiss flag of the League of Nations has been turned into a swastika (illus. 66). It comes as no surprise that Heartfield was forced to flee Nazi Germany in 1933, and he continued his

66 John Heartfield, *The Meaning of Geneva*, cover of *Arbeiter Illustrierte Zeitung*, 27 November 1932.

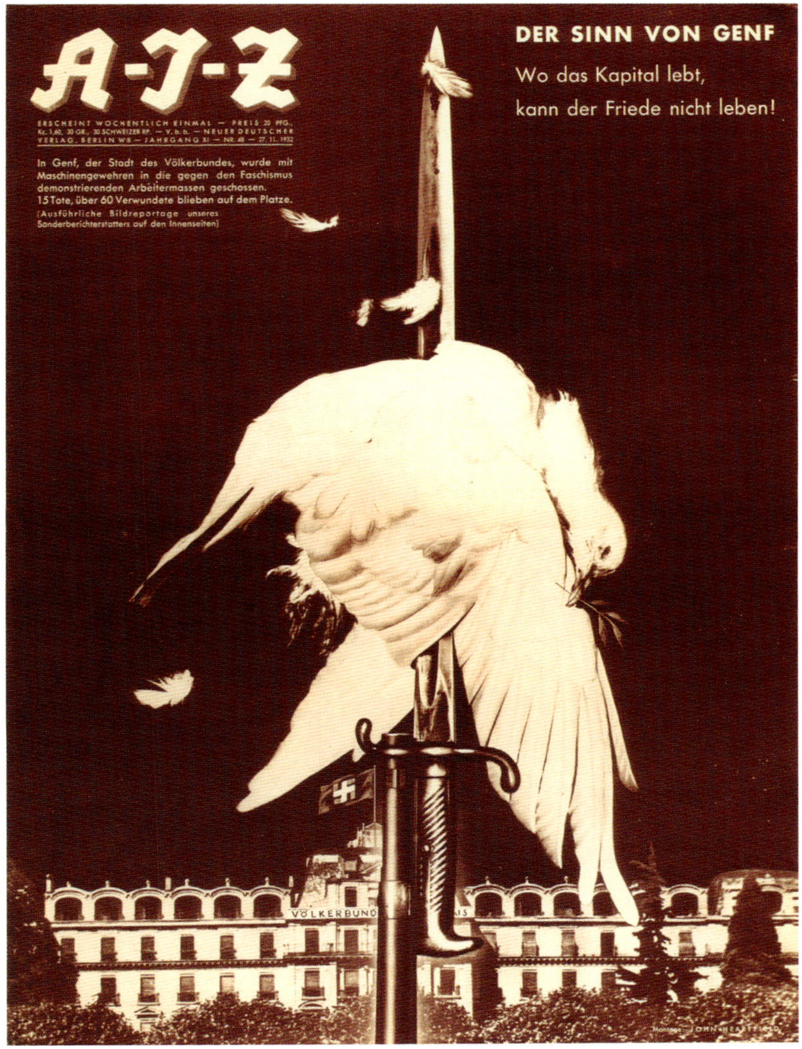

attacks from Czechoslovakia with photomontages protesting the Spanish Civil War in 1936 and the Japanese attack on Manchuria in 1937. He was a committed pacifist until his death in 1968, and his photomontage of the impaled dove was reissued as a poster in 1960 during the Vietnam War, under the slogan 'Niemals Wieder!' (Never Again!) Collectively, Heartfield's

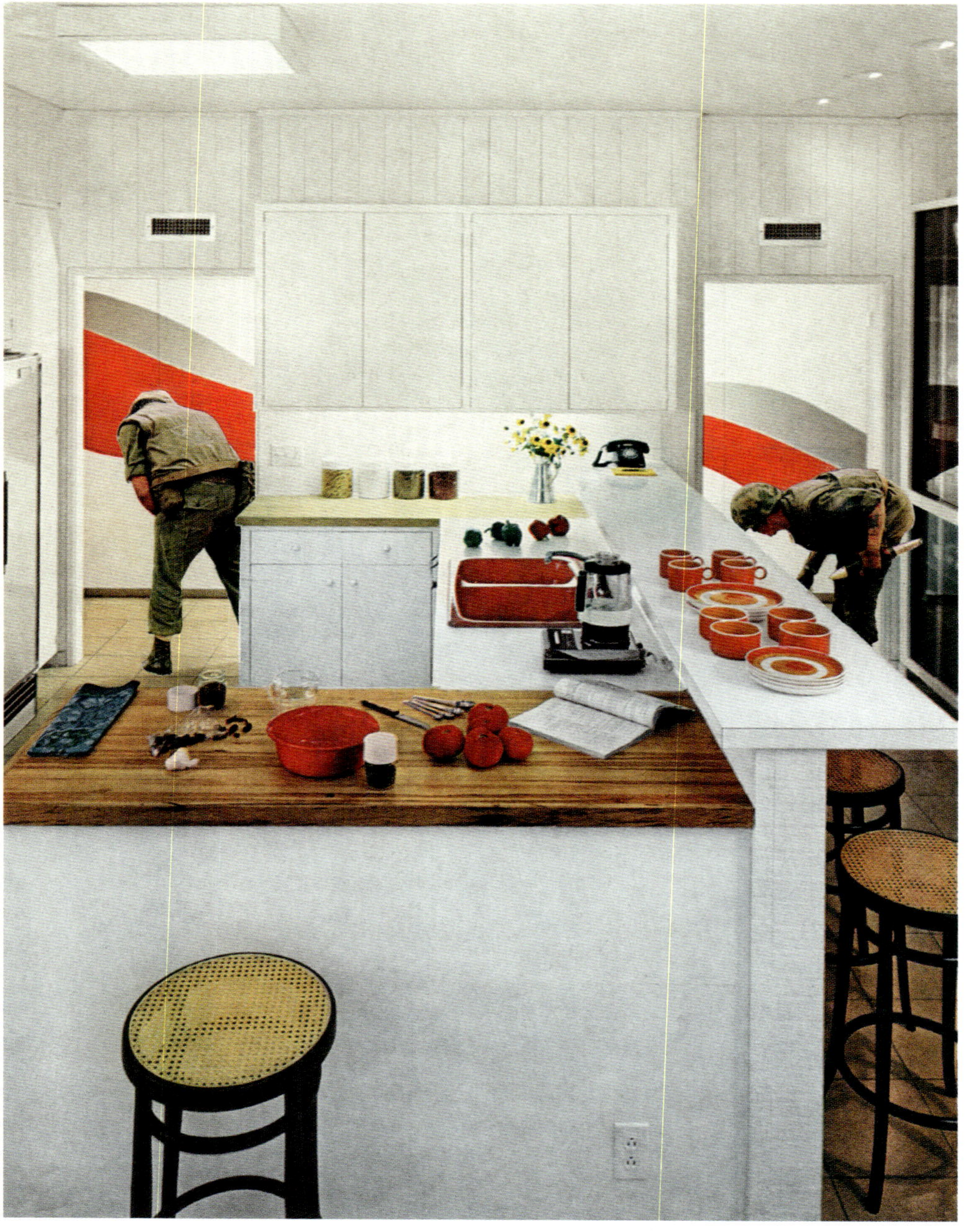

67 Martha Rosler, *Red Stripe Kitchen*, from *House Beautiful: Bringing the War Home*, *c.* 1967–72, photomontage.

works remain some of the most arresting visual statements against war, of any epoch.

Heartfield died in 1968, but his trenchant political critique has been continued by numerous other activists. During the Vietnam War, the U.S. theorist and artist Martha Rosler made a series of photomontages entitled *House Beautiful: Bringing the War Home* (*c.* 1967–72). Although Rosler's work is now held in public art collections, the photomontages were originally made for flyers at anti-war protests and underground newspapers, circulated by activists to catalyse opposition to U.S. intervention in Vietnam. *Red Stripe Kitchen* (illus. 67) depicts a gleaming white fitted kitchen, fashionably kitted out with Thonet barstools, red sink and electric blender: the kind of 'dream home' advertised in consumer magazines for women. The domestic scene is disrupted, however, by the incongruous presence of two soldiers clad in flak jackets, helmets and khaki trousers of the type worn by U.S. soldiers serving in Vietnam. Armed with what appears to be a rocket-propelled grenade, the men peer round the doorways. In the pristine white kitchen, their tropical-issue camouflage attire makes them conspicuous rather than invisible.

Rosler's image is unsettling: what place do these soldiers have in the home, bringing with them the threat of violence? How have they been permitted to invade this safe space? Rosler repurposes images found in *Life* magazine, that bastion of all-American values and classic war photography, with dark humour and surrealism. *House Beautiful* offers a many-layered commentary on the Vietnam War, suggesting how the far-flung conflict, which became known as 'the living room war', entered the home via nightly television news reports. The aspirational kitchen might be interpreted as symbolizing U.S. consumerism and bourgeois capitalism, the survival of which is at stake in the proxy war between East and West. Perhaps, also, Rosler questions those who perceive the sphere of women to be apolitical and unconnected to the masculine arena of war. *House Beautiful*'s visual collapse of boundaries between home and front suggests the complicity of *all* citizens in their nation's military and foreign policy, and their responsibility for deaths in far-off places.[39]

Women have organized collectively to directly criticize their governments' support of war, sometimes as part of broader attacks on patriarchy and on militarism as a masculine endeavour. At the height of the Cold War in the 1980s, a group of women achieved international notoriety when they protested the site of U.S. cruise missiles at the British air base RAF Greenham Common. Photographic images were a fundamental tool in the ongoing battle between the encamped women and the establishment. Protestors organized photogenic spectacles, such as women linking hands around the 9½-kilometre (6-mi.) perimeter fence or dancing on a missile silo, which were documented by sympathetic photographers including Raissa Page, Jenny Matthews and Maggie Murray, all of whom were members of the women's photography agency Format.[40] As well as featuring in the press, images were disseminated via flyers, pamphlets and posters as part of a wider visual culture of protest that included illustrations and hand-sewn banners. Photographs were also used as objects in demonstrations. An image made by Danish photographer Sigrid Møller, working on behalf of the Women's International League of Peace and Freedom, shows how protesters attached family photographs, along with other personal objects such as ribbons and

68 Sigrid Møller, perimeter fence at Greenham Common Women's Peace Camp, from *A Day in December 82*, 1982, colour 35mm transparency.

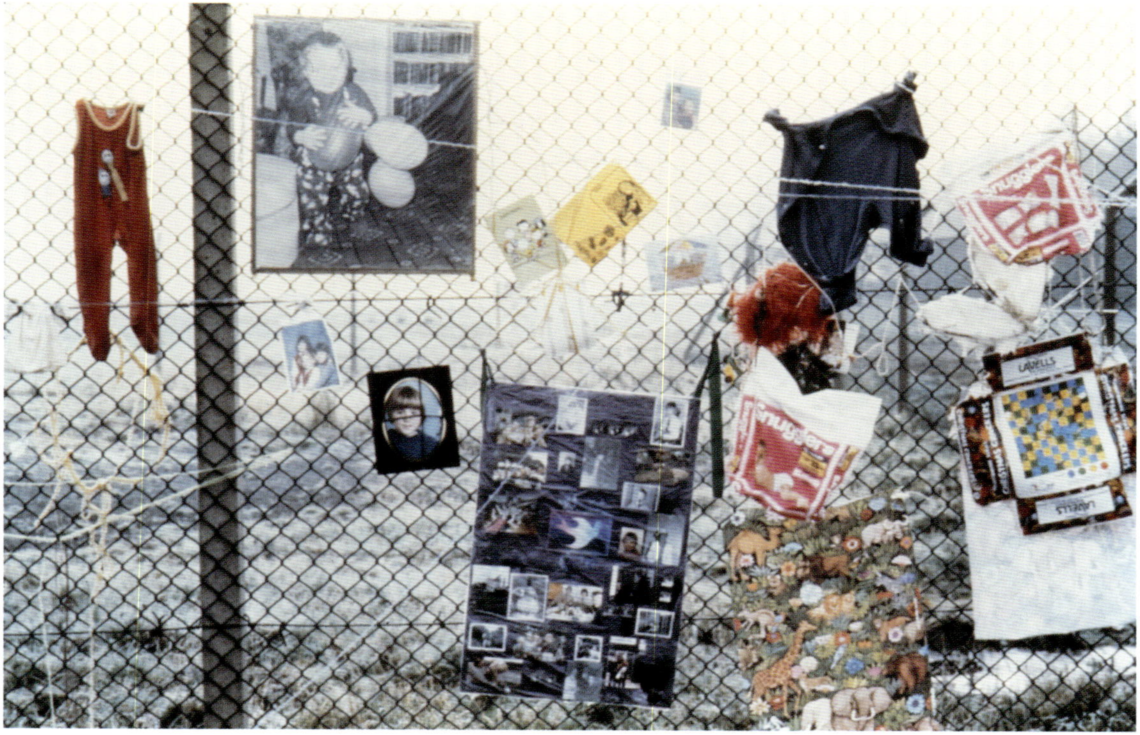

baby clothes, to the fence around the base (illus. 68). The pathos of these small, everyday objects conveyed just what was at stake in supporting a potentially devastating nuclear war. While the influence of the Greenham Common on military policy is debatable, they achieved worldwide recognition for their collective anti-war protest.

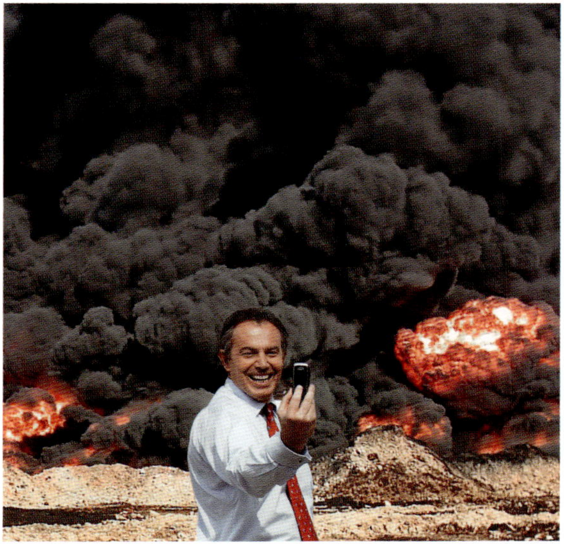

69 kennardphillipps, *Photo Op*, 2005, pigment print on paper.

In the twenty-first century photographic technology has become increasingly sophisticated, with possibilities for digital manipulation and seamless photomontages. The British political artist Peter Kennard, who has produced handmade photomontages since the 1980s, has subsequently collaborated with digital artist Cat Phillipps. Working together as kennardphillipps, the duo produced *Photo Op* (2005), a startling image of British prime minister Tony Blair apparently taking a selfie in front of an explosion during the Iraq War (illus. 69).[41] The image was first circulated as a newsprint handout at the G8 Summit protest in 2005, and subsequently proliferated in a range of contexts including newspapers, pamphlets, exhibitions and even on the side of a London bus. The image became, according to *The Guardian* critic Jonathan Jones, 'the definitive work of art about the war' and, quite possibly, how many people will remember Blair's role.[42]

Laudable as they are, it must be remembered that the photographic protests of kennardphillips et al. were permitted to circulate freely in the liberal democracies of the U.S. and the UK. However imperfect these polities may have been in other respects, they did not (as the Nazi regime did) suppress critique by citizens: the corrosive *Photo Op* image of Tony Blair was even acquired by the state-funded Imperial War Museum as part of its national collection. Challenges to military regimes and political dictatorships elsewhere, however, have been far more perilous. Photographic resistance may lead to persecution, exile, torture, execution or 'disappearance', yet people have repeatedly risked their lives to expose clandestine violence and seek justice or retribution. Struggles over visibility and concealment are explored next.

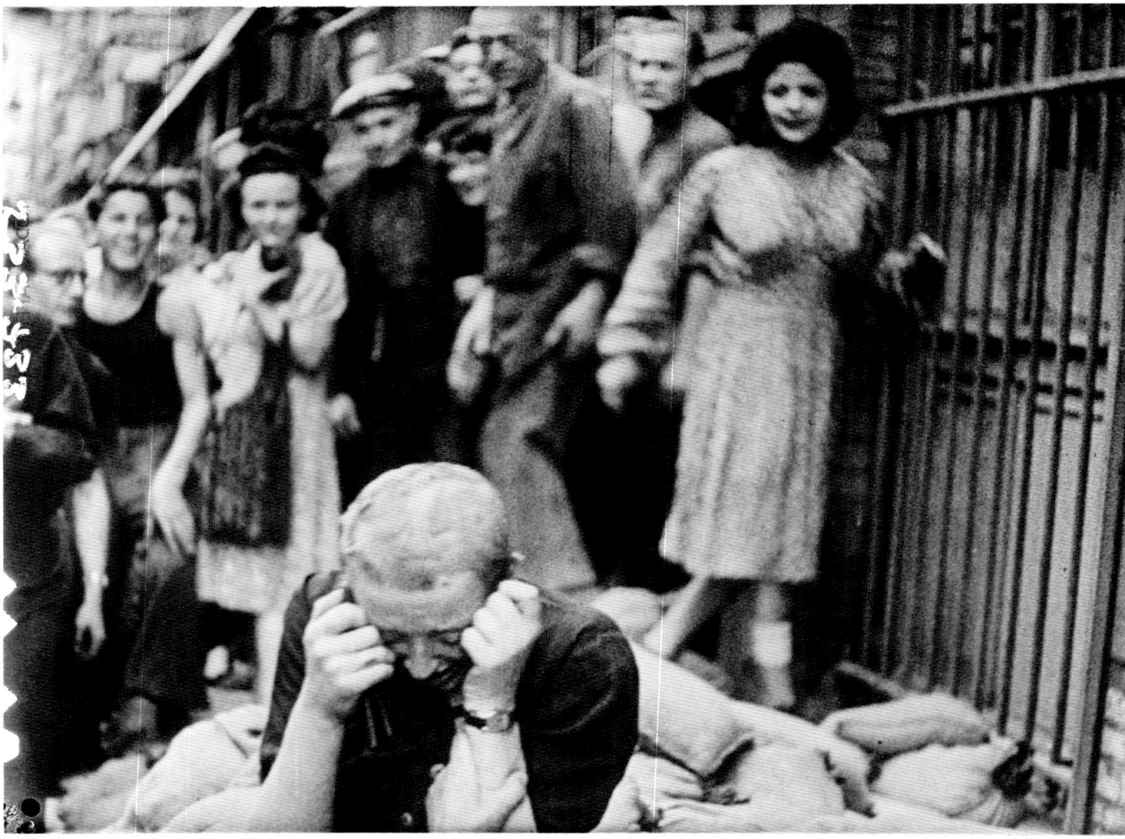

four

Secrets and Exposures

The language of truth and justice is suffused with metaphors of illumination and making visible: crimes are brought to light, perpetrators exposed, the truth revealed. Repression and war crimes are, by contrast, characterized by darkness and concealment: covert action, forced disappearance, black ops. Photography has a peculiar affinity with the notion of 'exposure', in both its sense as the revelation of something secret and potentially damaging, and the action – at least, in the pre-digital era – of exposing a photographic film to light, to be indelibly imprinted by an event.

As demonstrated by many of the cases in this book, the assumption of photography as a truthful and objective medium is problematic. And yet, the political value of photography's evidentiary power may, under specific circumstances, be made useful. Its assumed status as an unmediated eyewitness has meant that it has repeatedly been called upon to disclose crimes. 'Should not every photographer', pondered the German Jewish intellectual Walter Benjamin, 'expose guilt on his pictures and identify the guilty?'[1] Benjamin, who took his own life in 1940 rather than be captured by the Nazis, was speaking metaphorically, but he would surely have appreciated those photographers who took great risks to make clandestine records in the hope of one day exposing their persecutors. Some images have erupted into public view almost immediately, while others have remained in obscurity until dislodged, sometimes by chance, years after conflict has officially ended.

Conversely, perpetrators may strategically hide aspects of war from view, or use photographs to repress, humiliate or violate. Covert activities thrive in an atmosphere of political paranoia, driven by ideological, religious or ethnic hostilities, and are marked by illegal practices, human rights abuses and war

70 Unidentified *tondue* (shorn woman), France, c. 1944, black-and-white snapshot.

crimes. Certain conflicts have been especially characterized by secrecy and exposure: the Cold War; the 'dirty' wars of Latin American dictatorships; state-sponsored violence in Africa and East Asia; the ongoing War on Terror; civil conflicts, guerrilla insurgencies and paramilitary operations. As the following discussion shows, however, the suppression of abhorrent acts – and their subsequent exposure via photographs – has been carried out in a wide range of conflicts, and by factions that might otherwise be considered morally 'right'.

While many of the images in this book are troubling, the ones that feature in this chapter are particularly disturbing, and many are sickening. The decision to include these images has been no easy matter, and risks inflicting a kind of double violence on victims already degraded by the original pictures. However, it would be negligent to shy away from this unpalatable but fundamental aspect of photography's relationship with war.

Espionage and Surveillance of Civilians

Photography's facility for rapidly recording and transmitting information has made it an indispensable tool for intelligence gathering. The military use of miniaturized photographic images is thought to date to the American Civil War, when intelligence agents are reported to have transported microphotographs of despatches in their coat buttons.[2] Carrier pigeons were the transporters of choice in the 1871 Siege of Paris (see illus. 41); during the Russo-Japanese War (1904–6), couriers of the Japanese Imperial Army were reported to have carried microphotographic messages in their nostrils, under their fingernails or in ivory phials they had swallowed.[3] Microphotography was dramatically improved by Emmanuel Goldberg, who publicized his *Mikrat* (microdot) apparatus in 1925, which could transform fifty lines of print to the height of 0.1 mm. A Russian Jew, Goldberg fled Germany in 1933, but his invention was subsequently used in the Second World War and the Cold War, with microdots applied to documents in the guise of punctuation, or inserted into the edges of postcards, among other methods.[4]

Cameras were also miniaturized. The Latvian German scientist Walter Zapp invented the diminutive 'sub-miniature' Minox camera, first manufactured in Riga in 1938. Although originally intended as a slimline camera marketed to women, its potential for espionage was grasped by intelligence agencies.[5] Sub-miniature cameras were used to copy classified documents and maps in the Cold War, when information of all kinds – but particularly pertaining to weapons development and the 'space race' – became crucial to the East–West military stand-off.[6] Soviet spy Christopher John Boyce and East German double agent Heinz Felfe, among others, were found in possession of Minox equipment. The F21, developed in the late 1940s by the Soviet Intelligence Agencies, became standard-issue for the East German Stasi secret police. One application comprised a camera lens hidden behind a button, while the wearer squeezed a shutter cable from their coat pocket (illus. 71). The F21 was hidden in all kinds of everyday objects – Thermos flasks, radios, umbrellas, cigarette packets – and could be concealed behind belt buckles or even in bras.[7] By the end of the 1960s, sub-miniature cameras had assumed popular currency, with a Minox III featuring in the James Bond film *On Her Majesty's Secret Service* (1969).

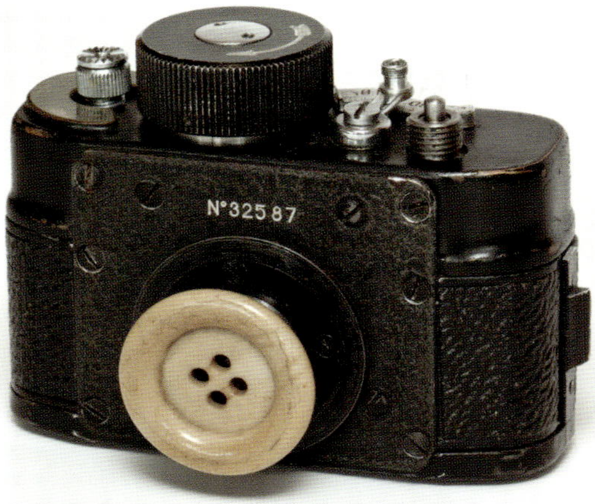

71 Minox F21 buttonhole spy camera, 1960–70s.

Covert photography has often been a means of identifying and monitoring real or alleged terrorists, spies and dissidents, particularly in ideological conflicts where the enemy may be so well integrated into society as to be virtually undetectable. During the Cold War, the Stasi secret police conducted prolonged and widespread surveillance on citizens suspected of 'subversion'. At least 1.7 million photographic records have survived, conserved in the BSTU (Bundesbeauftragte für die Stasi-Unterlagen, Stasi Records Agency) in Germany.[8] The artist Simon Menner spent two years researching and categorizing photographs held in the archive, constructing typologies such as 'Spies Photograph Spies' and 'Teenager Rooms'.[9] A series of grainy black-and-white images entitled *Mailbox Surveillance* monitors letters despatched by citizens including a pair of innocent-looking ladies with a toddler in a pram (illus. 72). Another series comprises Polaroids depicting the intimate details of homes before agents carried out secret searches, enabling items to be returned to their original state undetected, from the contents of a drinks cabinet to an unmade bed. These apparently innocuous scenarios are contaminated by a sinister air of suspicion, perhaps the prelude to arrest and interrogation.

Beyond the familiar East–West axis, the Cold War also fuelled a climate of ideological persecution in other regions, including Latin America. Under the aegis of counter-insurgency, the right-wing Ríos Montt dictatorship in Guatemala, covertly supported by the U.S. Central Intelligence Agency, conducted surveillance against its own citizens from around 1960 to 1996. In one of the continent's most startling rediscoveries of the so-called 'archives of terror', the records of the National Police were found in a semi-derelict building in Guatemala City in 2005.[10] Among the huge heaps of mouldering documents were innumerable ID pictures, used to identify and target citizens for abduction, unlawful imprisonment, torture and execution. In the face of official denials, the survival of these photographs attests to the systematic and repressive surveillance carried out by the state, with implications for ongoing processes of truth and justice.

During the War on Terror, the international military campaign launched by the U.S. after the terrorist attacks of 9/11, surveillance of citizens has been a controversial strategy. Proponents defend it as necessary to ensure the

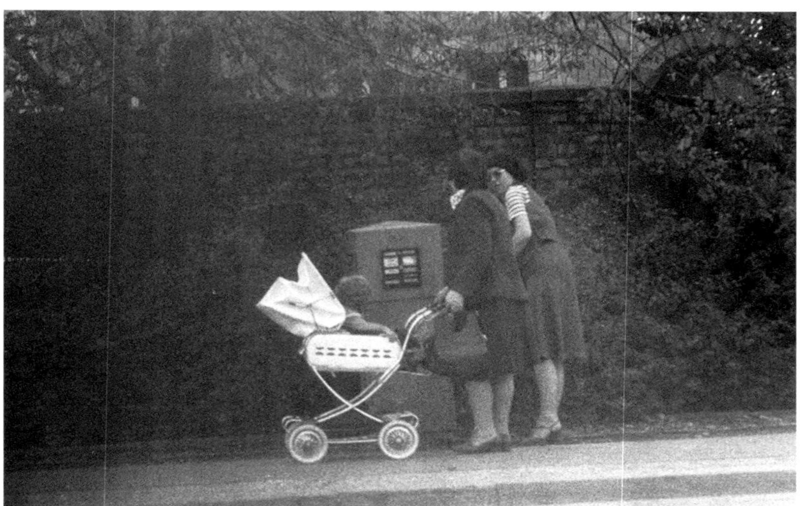

72 Simon Menner, from the series *Mailbox Surveillance*, c. 2013, silver gelatin print from the Bundesbeauftragte für die Stasi-Unterlagen (Stasi Records Agency), *c*. 1960s.

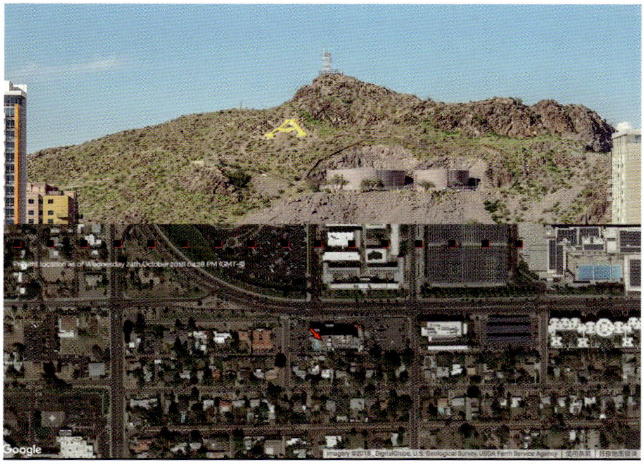

73 Hasan Elahi, *Tracking Transience*, 2005–present, colour digital images from website.

nation's security; opponents question the legality and ethics of monitoring suspects via mobile phone networks, Internet footprints and CCTV cameras in public spaces. In the wake of 9/11, Bangladesh-born, U.S.-based artist Hasan Elahi found himself the erroneous subject of a six-month-long investigation by the FBI. In response, he began to voluntarily monitor himself through photographs detailing the activities of his daily life. Begun in 2003, his ongoing web-based project *Tracking Transience* uses GPS (Global Positioning Satellite) technology in tandem with Google Maps to provide a live visual representation of the artist's movements, accompanied by photographs made during his travels (illus. 73). The print installation *Thousand Little Brothers* (2014) features images that Elahi sent unbidden every week to the FBI: a deluge of dinner plates, toilet seats and hotel beds. Elahi's actions function both as critique of technologies of surveillance and as a failsafe alibi in the event of future accusations.[11]

Perpetrator Photography

Surveillance photographs tend to be characterized by latent rather than explicit violence. By contrast, so-called 'trophy' photographs depict in grisly detail the suffering of enemy victims, perhaps made to be gloated over afterwards. Although often clandestine, trophy photos occasionally surface to trouble the conventional photographic discourse of war. At the turn of the nineteenth century, graphic images allegedly depicting atrocities committed in the region of Ottoman Macedonia appeared in popular illustrated journals in Europe. The cover of the 27 February 1903 edition of *La Vie illustrée* (illus. 74) features an extraordinary image of a group of armed soldiers, wearing the fez of the Ottoman Army, posed around a decorative plinth on which are arranged three heads. The caption declares, 'Turkish atrocities in Macedonia – before the lens. Group of Turkish soldiers pose at the photographer's

Numéro 228 27 Février 1903

LA VIE ILLUSTRÉE

LES ATROCITÉS TURQUES EN MACÉDOINE — DEVANT L'OBJECTIF

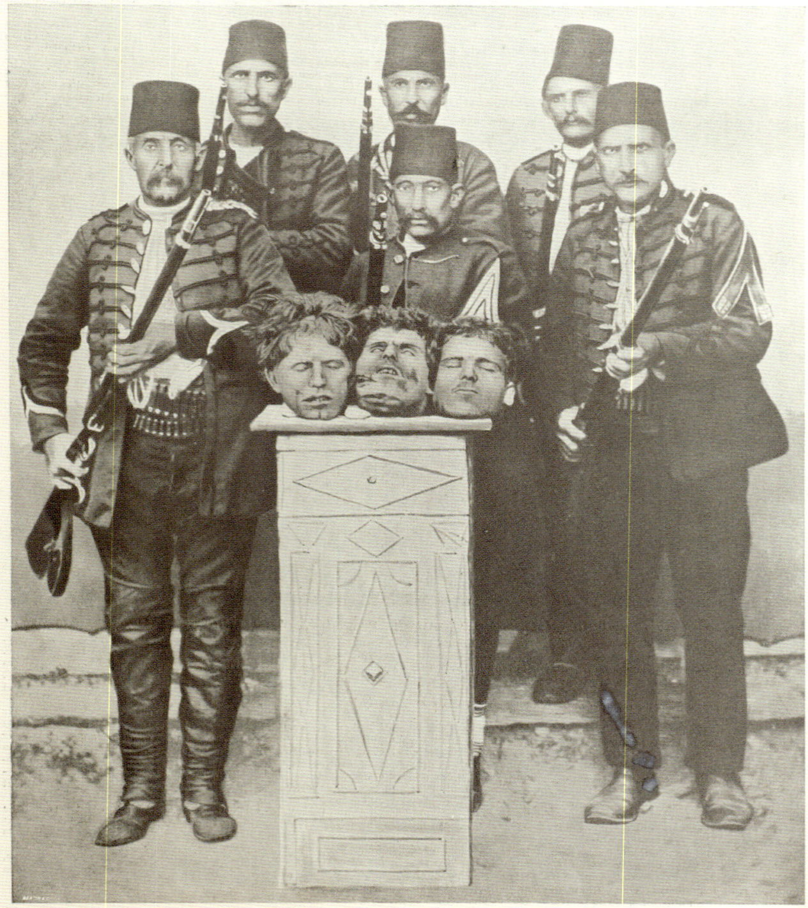

GROUPE DE SOLDATS TURCS POSANT, CHEZ LE PHOTOGRAPHE, AVEC LES TÊTES DE LEURS VICTIMES [studio] with the heads of their victims.' What is perhaps most startling is the care that has been taken to create this formal arrangement: a symmetrical tableau of rifles, fezzes and severed heads displayed as a particularly gruesome kind of still-life. As Turkish historian Edhem Eldem has asserted, it is impossible to determine the authenticity of the images, which were vehemently denied by the Ottoman authorities as 'evil fabrications'. While the severed heads are real enough, the scene itself may have been staged, or perhaps made at an earlier date in equally horrible, but not necessarily politically motivated, circumstances.[12] The case illustrates both the difficulties in authenticating 'trophy' photographs and the degree to which their impact depends on the context in which they are circulated: it is likely that these particular images were disseminated by Macedonian freedom fighters as proof of assumed Turkish 'barbarism', functioning as propaganda to galvanize European opposition to the Ottoman Empire.[13]

According to a subsequent article in the *Berliner Illustrirte Zeitung* of 15 March 1903, the Turkish soldiers brazenly carried sacks of severed heads to the nearest professional photographer to have their portrait made.[14] As the popularity of snapshot cameras grew, however, soldiers could easily make photos themselves.[15] The opprobrium surrounding this activity has meant, unsurprisingly, that such photographs remain clandestine or shared within closed circles, hidden in private photo albums or – in the digital age – underground 'dark' websites. Occasionally such images do emerge, often with spectacular and controversial effects. The disclosure of photos made by military guards at Iraq's Abu Ghraib prison in 2004, discussed later in this chapter, is perhaps the most prominent scandal of the digital era (see illus. 75). Equally troubling, but much older, were the images featured in the exhibition *Vernichtungskrieg: Verbrechen der Wehrmacht 1941 bis 1944* (War of Extermination: Crimes of the Wehrmacht, 1941 to 1944), shown in numerous venues in Germany and Austria between 1995 and 2000.[16] Conceived and researched by the Hamburg Institute for Social Research, the exhibition included several hundred snapshots thought to have been made by 'ordinary' German soldiers during the Holocaust that showed the victimization of Jews, civilians hanging from gibbets and corpses piled in mass graves. While atrocities committed by Nazi officials had been acknowledged, the implication of much more widespread barbarity throughout the combined armed forces

74 Unidentified photographer, 'Turkish Atrocities in Macedonia – Before the Lens', cover of *La Vie illustrée*, 27 February 1903.

traumatized the nation, and elicited a ferocious response. Petra Bopp, the exhibition's organizer, noted that some veterans rejected the photographs wholesale, while others declared that the reality had been very much worse.[17] As criticism from politicians, veterans and academics escalated, the provenance and identification of some of the photographs was questioned, and ultimately the entire enterprise was undermined. The exhibition was recalled, and a more anodyne version substituted. The episode illustrates not only the challenges of curating 'difficult' historical pictures, but of debating their implications in public.

In some instances of perpetrator photography, the victims are alive and cognisant of being depicted. Indeed, abuse or violence may be staged precisely to form a visual spectacle, and the photographic act wielded as a further layer of humiliation beyond the actual assault. During the Épuration sauvage (Wild Purge) in France following the liberation of 1944, women accused of collaborating with the occupying Nazi forces were rounded up and ritually humiliated. Although not all women were accused of sexual collaboration, the so-called *femmes tondues* ('shorn women') were subject to a particularly gendered form of violence, being publicly shaved, stripped, paraded naked or daubed with lipstick swastikas. Jeering onlookers photographed the women's shame and terror.[18]

The best-known professionally made image of a 'shorn woman' was made by Robert Capa and depicts a crowd flanking a *tondue* holding a baby in her arms as she walks down a street in Chartres. Capa's skilfully composed image is unusual in affording the woman some dignity, even sympathy: although shorn, she remains clothed and cradles her baby in her arms in an expression of maternal concern, while the crowd appears controlled. Other records made by local photographers or onlookers with snapshot cameras convey more chaotic and vitriolic displays of public punishment. An image by an unidentified photographer shows a sobbing woman seated on a pile of sandbags, her head shaved and her fists clutched to her face (illus. 70). A hand grips her wrist, probably keeping her still for the photographer, while in the background a group of onlookers (some of whom appear to carry cameras) enjoy the spectacle. Images of *tondues* appeared immediately in the national and international press, but have been less present in subsequent official narratives of France's liberation. François Boucher, curator of the 1944 exhibition *La Grande délivérance de Paris* for the Musée Carnavalet in Paris, put out an open call for photographs to be included in both the exhibition and its accompanying catalogue. However, he rejected images of *tondues*, presumably because they disturbed the patriotic narrative of resistance and heroism under occupation.[19] Few images are prominent in public collections, and have largely been ignored in recent digitization projects, testifying to the still-corrosive nature of these troubling images.

One of the most notorious instances of perpetrator photography in recent years concerns U.S. servicemen and women serving as guards at Abu Ghraib prison in Baghdad, Iraq. In 2004 photographs emerged in the U.S. media showing prisoners suffering a range of abuse including simulated electrocution, attacks by military dogs and assaults of a sexual and scatological nature. The digital snapshots made by the guards demonstrate a sadistic pleasure in the humiliations of the dehumanized prisoners, and in arranging them in poses staged specifically for the camera.[20] The guards' bizarre practice of piling the prisoners into pyramids of orange-suited or semi-naked bodies recalls the arrangement of trophy heads by the Turkish soldiers in Macedonia (see illus. 74). Like their predecessors, the guards at Abu Ghraib follow the photographic conventions of their day: in this case,

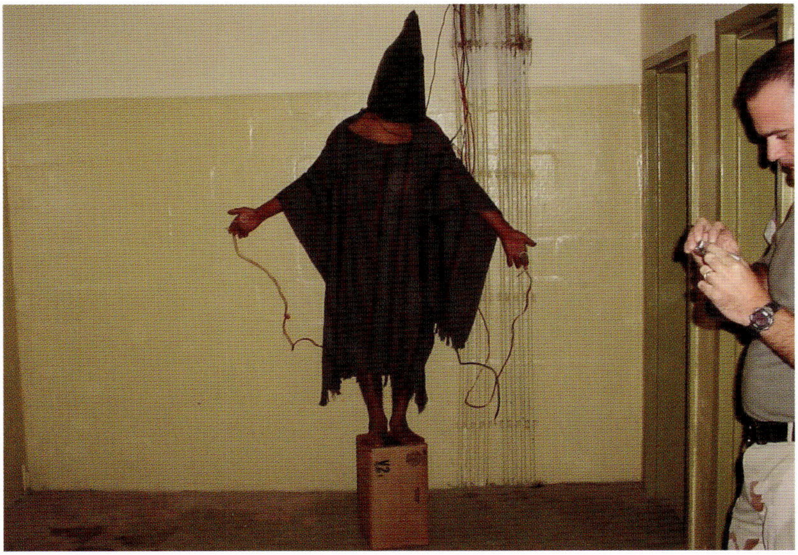

75 MP Spc. Sabrina Harman, prisoner being tortured at Abu Ghraib in the presence of Staff Sgt Ivan Frederick, 2003, digital colour photographic image.

snapshot aesthetics, mugging for the camera and making thumbs-up signs. The image that became the most iconic, made by Staff Sgt. Ivan Frederick, depicts a hooded man standing on a cardboard box with his arms apparently wired to electrodes. A companion image made at the same time by MP Spc. Sabrina Harman captures Staff Sgt. Frederick standing at the right-hand edge of the frame, examining what appears to be a digital camera, perhaps reviewing his images in playback mode (illus. 75).[21] The pathos and symbolism of the outstretched hooded figure, evoking both the garb of the Ku Klux Klan and the painterly tradition of the Crucifixion, contributed to it becoming a global icon of suffering.

The guards' images were made for their own personal use and not intended for wider circulation. They did not, however, stay private for long. After encountering the images on CDs, a colleague-turned-whistleblower alerted the Criminal Investigations Division, which brought the perpetrators to trial. The resulting classified 'Taguba Report' and associated images eventually found their way to the news network CBS and the pages of the *New Yorker* magazine.[22] International condemnation of U.S. brutality immediately followed, with the U.S. Army swiftly denouncing the perpetrators as a few 'bad apples', punished by military imprisonment. The episode led to rhetoric on restoring humane treatment of prisoners in the War on Terror, but while the photographs raised awareness for public debate, it is unclear if abusive practices have ended.

On other occasions, perpetrator photography has been officially sanctioned, employed by military regimes within a state-sponsored system of violence. Between 1975 and 1979, the regime of the Khmer Rouge in Cambodia conducted war against its own people in a genocidal campaign to remove 'subversives' and ethnic minorities. The mass scale of such an operation – between 300,000 and 500,000 people are thought to have been killed – necessitated a bureaucratic system in which photography was embedded.[23] Nhem Ein has since recounted his role as the chief of six photographers at the S-21 security prison in Phnom Penh.[24] Photographers were employed in the documentation unit to make mugshots of inmates as they entered the prison, to photograph those who died while in detention, and, in some cases, record victims after execution. One image depicts a boy naked to the waist, his arms behind his back (illus. 76). A label with the

76 Nhem Ein, prisoner at S-21 security prison, Phnom Penh, Cambodia, *c.* 1975–9, silver gelatin print made later by Photo Archive Group.

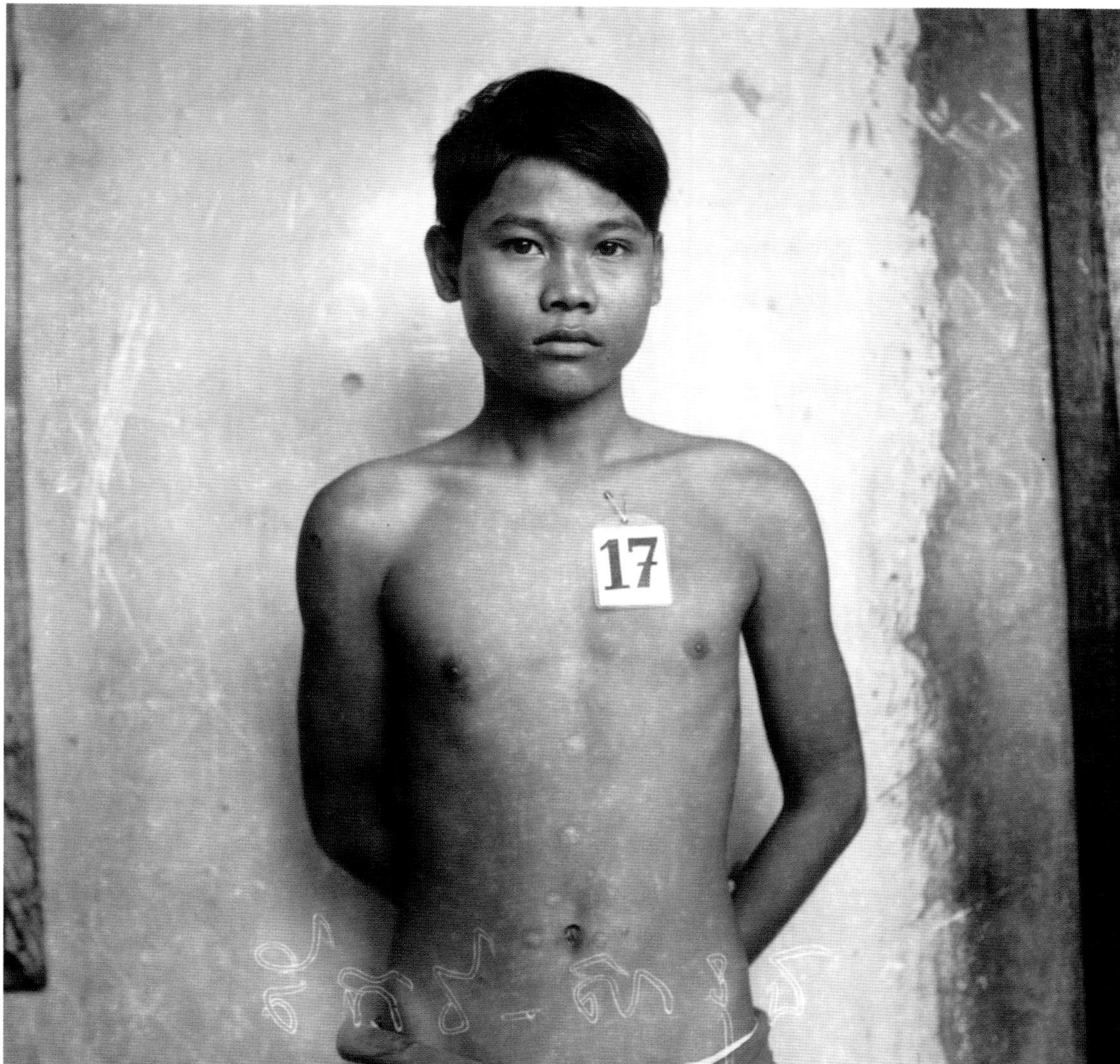
17

number 17, presumably correlating to the prisoner's dossier number, is crudely attached to his chest with what appears to be a safety pin.[25] The young man stares impassively towards the camera and, by proxy, to the present-day viewer of the image. The affective power of the image, like those pertaining to the Holocaust and other genocidal campaigns, is predicated on the poignancy of the subject impassively facing an unknown future, and the viewer's knowledge of the death that awaits.[26]

About 6,000 negatives have survived and are conserved in the Tuol Sleng Museum of Genocide in the former S-21 building.[27] These represent a fraction of those originally taken, and the majority of subjects remain unidentified. The comprehensiveness of the regime's record-keeping represents a colossal act of hubris: presumably it was never imagined that these state-sanctioned records would be spectacularly transformed into incriminating evidence. A selection is on permanent display to visitors in the museum, functioning as both a record of national trauma and a means to encourage onlookers to become 'messengers of peace'.[28]

Incarceration and Prison Photography

Wartime imprisonment is often overlooked by histories of war and war photography, perhaps because access to depicting experiences 'behind the wire' is by nature restricted, and perhaps because the assumed passive condition of captivity does not follow the script of the heroic soldier narrative. Prison photography is a hugely diverse practice, undertaken in a range of contexts and for divergent reasons. The images from S-21 and from Abu Ghraib, discussed above, were the products of perpetrators. At other times, however, prisoners have been more sympathetically portrayed by outsiders or have made photographs themselves.

Dorothea Lange is best known for her photographs made for the U.S. government agency FSA (Farm Security Administration) during the Great Depression of the 1930s. She later produced a compelling body of work on Japanese American internment during the Second World War.[29] Catalysed by the Japanese attack on Pearl Harbor on 7 December 1941, Order 9066 was the U.S. government's response to the so-called 'Japanese Problem' on the U.S.

West Coast. Lange was among a number of photographers commissioned in early 1942 by the War Relocation Authority to record the forced evacuation and incarceration of an estimated 120,000 persons of Japanese ancestry, many of whom were American citizens or relatives of U.S. soldiers. Lange's images convey the calm dignity of 'evacuees': a portrait of the Mochida family conveys the pathos of children labelled with luggage tags to avoid separation, while their father rests a reassuring hand on the shoulder of the youngest (illus. 77). Lange subsequently photographed 'assembly centres' where internees were processed, and the basic conditions of the principal camp at Manzanar, California. Lange was presumably hired on the basis of her compassionate images – such as the iconic *Migrant Mother* (1936) – that had previously documented government aid to dustbowl communities, but she found it impossible to support the racist agenda of Order 9066. Most of her images were deemed unsuitable for public relations purposes, and were impounded until after the war. The repressive treatment of Japanese Americans has been largely absent from mainstream narratives that foreground the U.S. as an upholder of freedom and democracy. Lange's

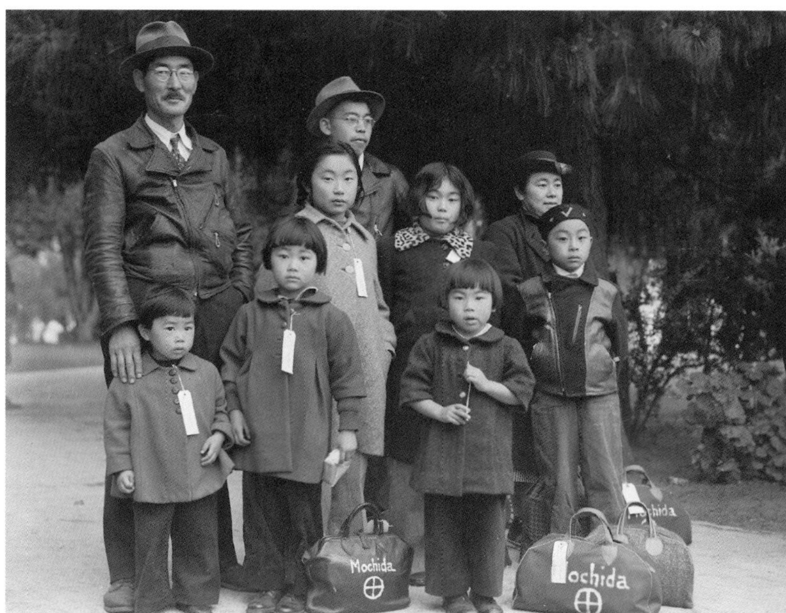

77 Dorothea Lange, members of the Mochida family awaiting evacuation bus (Hayward, California) 8 May 1942, silver gelatin print.

images were eventually made public in the exhibition *Executive Order 9066* at the Whitney Museum, New York, in 1972, but remain relatively little known.[30]

An insider's viewpoint of the Manzanar camp was made by Tōyō Miyatake, a professional photographer who had been working in Los Angeles's Little Tokyo before being incarcerated with his family in 1942.[31] Cameras were prohibited, but Miyatake managed to smuggle a lens into the camp and was eventually permitted to make photographs as the official camp photographer, on condition that a guard tripped the shutter.[32] Miyatake documented the humanity and resilience of his fellow inmates continuing their lives under great strain. A number of his photographs were reproduced in *Our World*, the 1943–4 yearbook of Manzanar High School (illus. 78).[33] Images of baseball games, weddings and Boy Scout meetings are punctuated by occasional glimpses of the barbed wire and watchtowers that surrounded the camp. The ambivalent publication serves an awkward dual purpose as both a statement of allegiance to all-American values and a historical record of a bitter segregation.

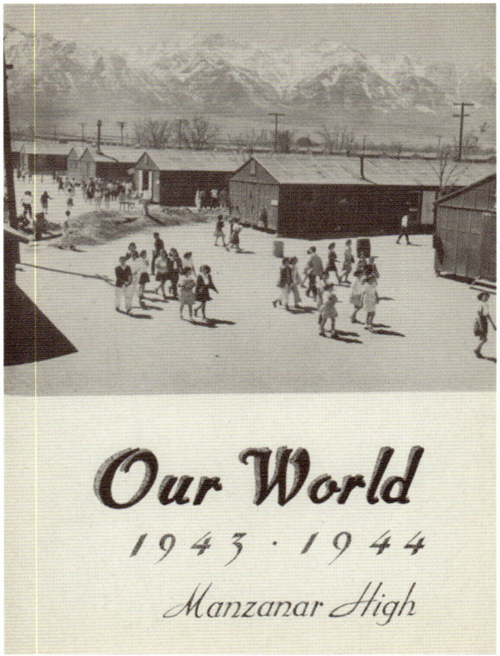

78 Tōyō Miyatake, cover of *Our World*, yearbook of Manzanar High School at Manzanar War Relocation Center, 1943–4.

In other cases, photography by inmates during their incarceration has not been possible, and post hoc approaches have been required to convey the experience. During Argentina's military dictatorship from 1976 to 1983, an estimated 30,000 citizens were 'disappeared' – that is, abducted, held without trial in secret interrogation centres, tortured and sometimes killed. One of those to survive their ordeal was the 22-year-old botanical student Paula Luttringer, who spent five months in a clandestine detention centre, where she gave birth to her first daughter. After being freed, Luttringer spent years in exile in Uruguay and France before returning to Argentina to take up photography. The nature of her experience, and of others like her, proved resistant to expression via conventional photojournalism. Instead, Luttringer conducted a kind of archaeology of terror, making black-and-white images of the remains of the sites of illegal incarceration, many of which were centrally located in towns and cities. She also collected testimonies of other female ex-prisoners, which

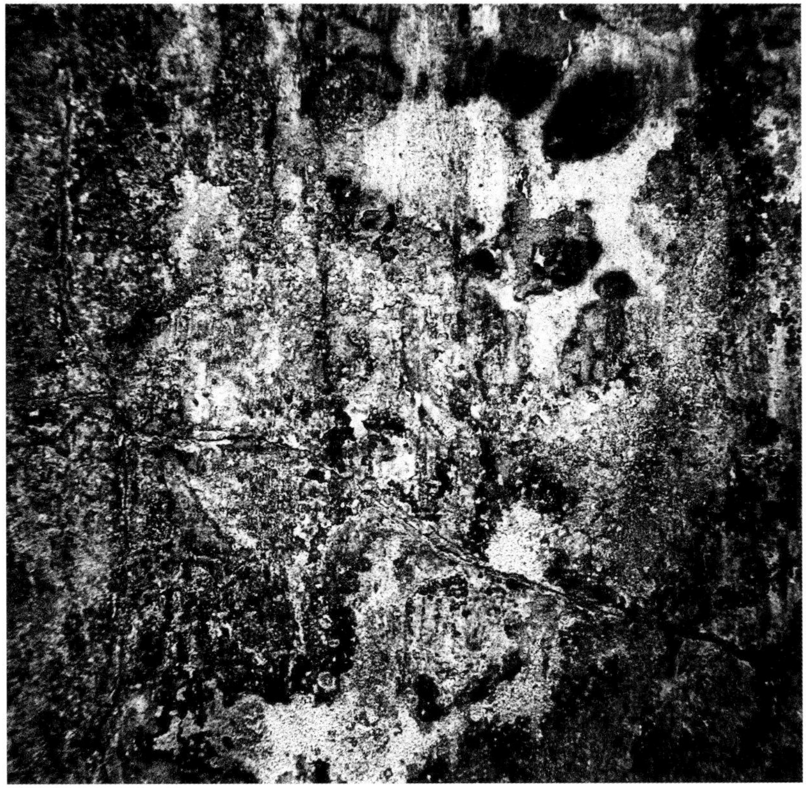

79 Paula Luttringer, from the series *El lamento de los muros* (The Wailing of the Walls), 2000–2010, carbon fresson print.

accompany the photographs.[34] An image of a grimy, indeterminate surface, perhaps a wall (illus. 79), is paired with the account of Isabel Cerruti, who was abducted on 7 June 1978 and held captive in the detention centre in Buenos Aires known as 'El Olimpo':

> Something strange used to happen at night, the screams of torture were different than those during the day. Even if the screams of torture are always the same they sound different at night. And it's also different when they come to get you at night. The noises and the screams are not with me always, but when I do remember them, it makes me very sad. I am paralyzed by those screams, I'm back in that time and place. As somebody once said – and I've given this some thought and I think it's

> right – although life goes on, although some of us were freed, you never get out of the pit.

Luttringer's dark and disorientating walls, stairs and ceilings place the viewer in the position of the prisoner herself, projected into the terrifying claustrophobic world of the cell. Entitled *El lamento de los muros* (The Wailing of the Walls), her series suggests that the traces of violence remain forever embedded in the fabric of buildings, as well as the minds and bodies of the victims. Words and images combine to powerful effect, helping reclaim the individual experiences and lives of 'the disappeared' in the face of official denials and impunity.

Covert Photography by Citizens and Resisters

Regimes have often suppressed the making of photographs, but citizens have been resourceful in their defiance. In the Nazi-occupied Netherlands the Journalist Decree of May 1941 imposed severe restrictions on photojournalists and professional photographers, yet they took great risks to record life under occupation. The clandestine group De Ondergedoken Camera (The Underground Camera) concealed cameras in briefcases and shopping bags, or behind coats and newspapers, in order to photograph in the street unobserved.[35] One of their members, Emmy Andriesse, was active in the resistance movement, producing photographs for forged documents produced by the Persoons Bewijzen Centrale (Centre for Identity Papers). Andriesse found herself in need of such papers herself when, in 1943, she was forced to wear a yellow star indicating her Jewish identity. After a year in hiding, she emerged with false papers and, using a concealed camera, photographed Amsterdam during the terrible *Hongerwinter* (Hunger Winter) of 1944–5. The German blockade of food supplies meant that official rations fell to a few hundred calories a day, and an estimated 20,000 people died of starvation. Andriesse's images register the pain of famished children forlornly waiting for food, most notably in an image entitled *Boy with a Pan* (1944) (illus. 80). With the help of Allied photographers, images by De Ondergedoken Camera found their way out of the country, and their

80 Emmy Andriesse, *Boy with a Pan*, c. 1944–5, silver gelatin print.

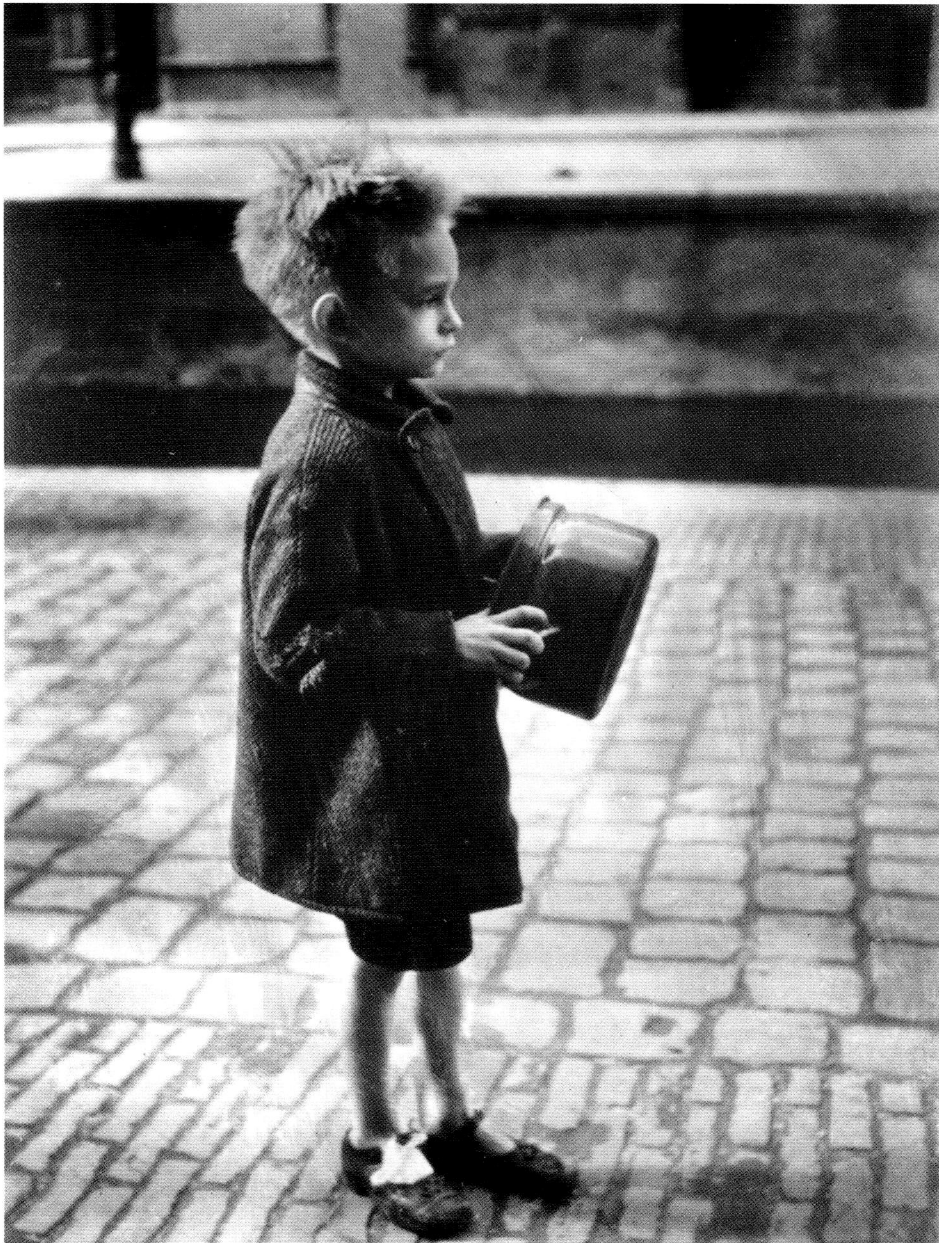

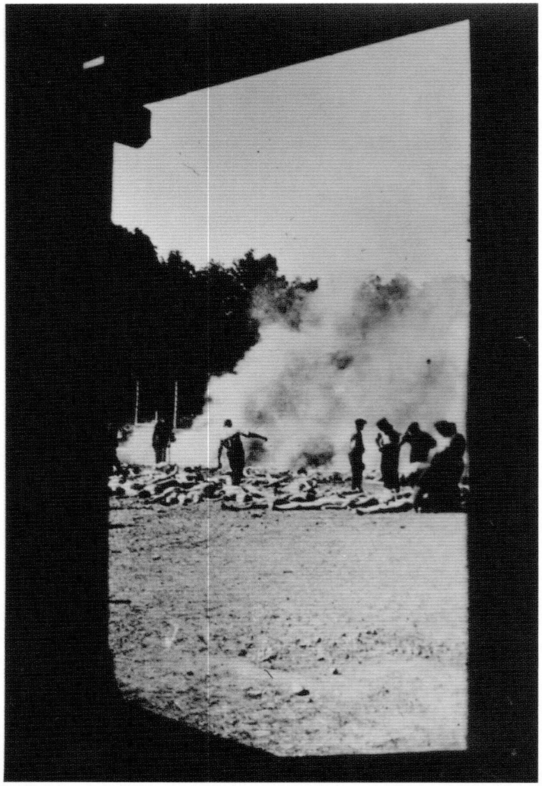

81 'Alex' (attrib.), cremation of bodies outside gas chambers at Auschwitz-Birkenau concentration camp, 1944, image from black-and-white negative.

work was subsequently published in the 1947 book *Amsterdam: Tijdens den Hongerwinter* (Amsterdam: During the Hungerwinter), helping preserve the collective memory of a national catastrophe.[36]

The mechanics of Hitler's 'Final Solution' to eliminate enemies of the Third Reich – Jews, 'subversives', Romany people, homosexuals and other non-conformists – culminated in a system of concentration camps where murder was committed on an industrial scale. There are very few extant photographs that record the horrors of the gas chambers in practice. Some were taken by an inmate known as 'Alex', thought to be Alberto Errera, a Greek-Jewish officer and member of the anti-Nazi resistance who was forced to work in a *Sonderkommando* (Special Unit) in the Auschwitz-Birkenau camp in 1944.[37] A camera was reputedly smuggled into the camp, hidden first in a bucket and later buried in the ground. At great risk, four images were made of women prisoners stripped in preparation for death, and the incineration of bodies immediately outside gas chambers (illus. 81). The film was smuggled out of the camp in a toothpaste tube and distributed by the Polish resistance. Although Errera died in Auschwitz in 1944, the images were subsequently reproduced by the Commission for the Investigation of German-Nazi Crimes in Oświęcim (Auschwitz), and remain indelible evidence of the Holocaust.[38]

Citizens have bravely used photography to resist regimes in numerous situations of repression. In South Africa during Apartheid of the 1970s and '80s, Peter Magubane documented the struggles from the perspective of a Black photographer, continuing his work despite repeated threats, assaults and periods of unlawful solitary confinement. Like the members of De Ondergedoken Camera, Magubane found inventive ways to conceal his true purpose, once hiding a camera in a hollowed-out loaf of bread.[39] A powerful image made during the student uprising in Soweto on 16 June 1976 shows two women approaching the photographer with the specific purpose of having him record police brutality. The face of one of the women is contorted in a scream; she supports her companion who pulls up her sweatshirt to reveal a bleeding stab wound in both a literal and metaphorical exposure of brutality. A committed activist, Magubane disseminated his photographs in hard-hitting publications including *Magubane's South Africa* (New York, 1978), *Soweto Speaks* (Johannesburg, 1979) and *The Fruit of Fear: June 16* (Braamfontein, 1986), helping convey the injustice of the regime under Apartheid.[40]

The Politics of Exposure

The risks that photographers have taken to record the abuses and repressions of armed conflict attest to the collective faith placed in photography's capacity to expose wrongdoing. However, the dissemination and reception of images is always contingent upon the political context in which they may (or may not) circulate. In the final bitter years of the American Civil War (1861–5), photographs emerged that claimed to depict Union soldiers mistreated in a Confederate prison camp at Belle Isle, Richmond, Virginia. Made at the U.S.

General Hospital at Annapolis, Maryland, the images show extremely emaciated former prisoners-of-war, such as Private Jackson O. Brashears of Co. D, Indiana Mounted Infantry . Some images were reproduced as *cartes de visite* (illus. 82).[41] Others were converted into wood engravings and reproduced in periodicals including *Harper's Weekly* and *Frank Leslie's Illustrated Newspaper*, as well as pamphlets such as *Narrative of Privations and Sufferings of United States Officers and Soldiers while Prisoners of War in the Hands of the Rebel Authorities*, published by the U.S. Sanitary Commission in 1864.[42] The alleged proof of the Confederates' lack of regard for the 'rules' of war caused an outcry, providing persuasive evidence of the Southern states' barbarism. The Confederates decried the images as the work of propagandists, some claiming (in a not entirely commendable defence) that conditions were at least as bad in Union prisons.[43] Nonetheless, the images were disseminated with the approval of the Secretary of War, Edwin M. Stanton, to strengthen Northern determination and refuel the conflict as a moral crusade.

War atrocities were exposed on an unprecedented scale when the Nazi concentration camps were liberated at the end of the Second World War. Allied troops, accompanied by international photojournalists, witnessed appalling scenes at Bergen-Belsen, Dachau, Buchenwald and other camps. Among the high-profile cohort were *Life* photographer Margaret Bourke-White; *Vogue* correspondent Lee Miller; and future Magnum co-founder George Rodger. The presence of photojournalists was a deliberate ploy by the Allied authorities to ensure that the magnitude of the Nazi atrocities would be known, in part to assuage war-weariness and reaffirm the necessity of fighting and ongoing privations. Throughout the spring and summer of 1945, images erupted from the pages of national newspapers and mass-market illustrated magazines such as *Life* and *Picture Post*, bookshops displayed pictures in their windows, and tens of thousands of people visited public displays in the UK and U.S.[44]

The Holocaust images remain some of the most troubling in the history of war photography, and the shock of their first appearance cannot be underestimated. Indeed, as photography scholar Sharon Sliwinski has argued, the inexplicable horrors of the genocide meant that the images initially met with a lack of comprehension. Photographs of earlier Soviet liberations of Majdanek and Auschwitz-Birkenau in Poland had largely

82 Photographer working at U.S. General Hospital, Annapolis, Maryland, *A Richmond Prisoner*, c. 1864, *carte de visite*.

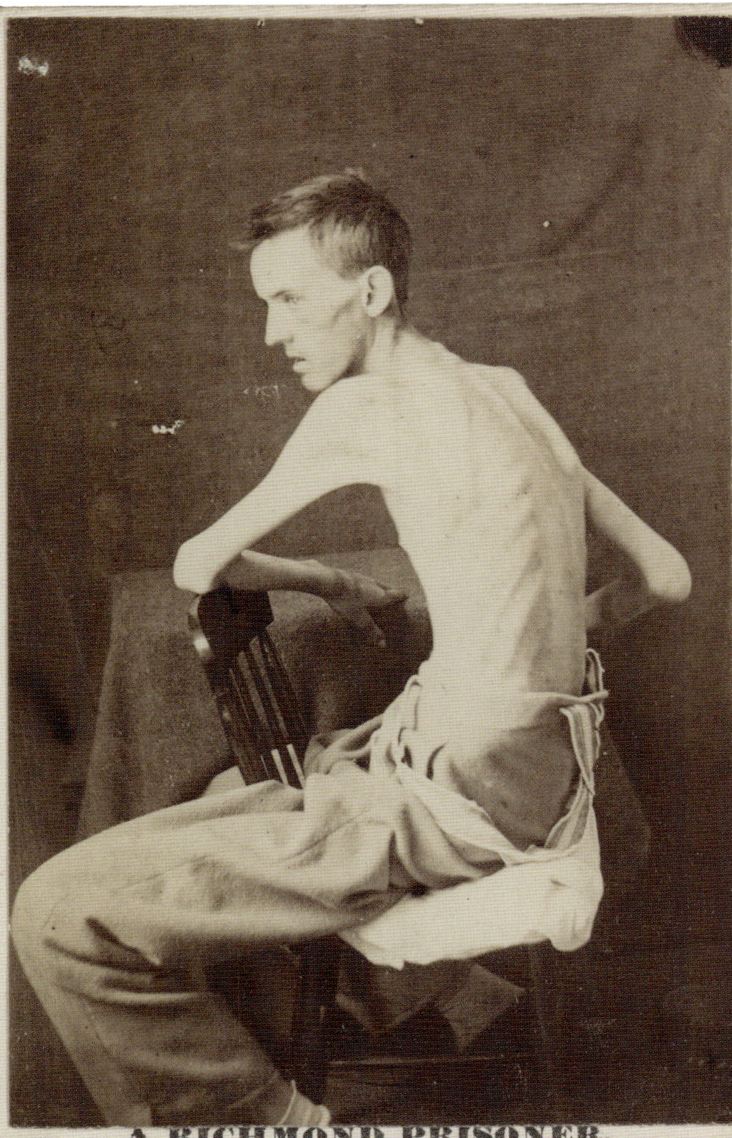

A RICHMOND PRISONER.

Entered according to Act of Congress, in the year 1864, by E. Wallace, in the Clerk's Office of the District Court of the Eastern District of Pennsylvania.

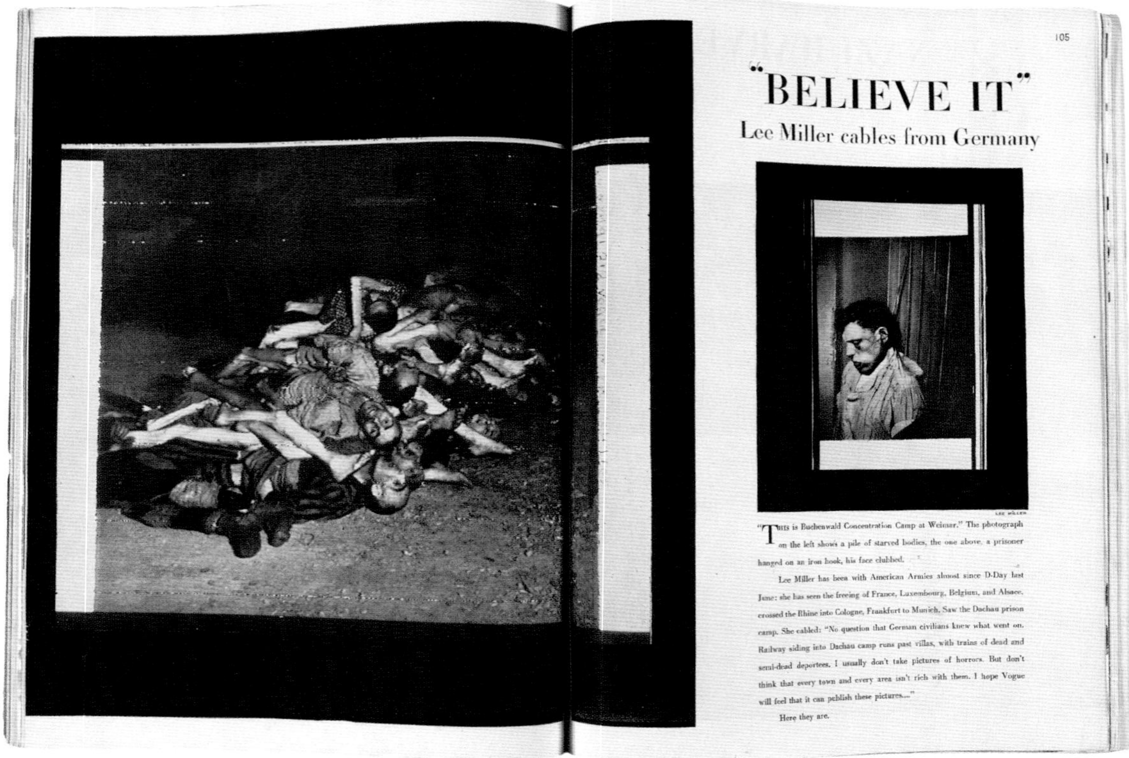

105

"BELIEVE IT"

Lee Miller cables from Germany

LEE MILLER

"This is Buchenwald Concentration Camp at Weimar." The photograph on the left shows a pile of starved bodies, the one above, a prisoner hanged on an iron hook, his face clubbed.

Lee Miller has been with American Armies almost since D-Day last June: she has seen the freeing of France, Luxembourg, Belgium, and Alsace, crossed the Rhine into Cologne, Frankfurt to Munich. Saw the Dachau prison camp. She cabled: "No question that German civilians knew what went on. Railway siding into Dachau camp runs past villas, with trains of dead and semi-dead deportees. I usually don't take pictures of horrors. But don't think that every town and every area isn't rich with them. I hope Vogue will feel that it can publish these pictures..."

Here they are.

83 Lee Miller, 'Believe It', double-page spread from American *Vogue*, June 1945.

been dismissed as Soviet propaganda, and some bewildered witnesses to Dachau and Buchenwald rejected the scenes as faked, produced by Allied agencies.[45] Yet as more atrocities emerged, and images continued to be published, photographs began to be accepted as compelling evidence for acts that could not otherwise be imagined. A photo-essay by Lee Miller, published in U.S. *Vogue*'s 1 June 1945 issue (illus. 83), depicts a tangle of naked and emaciated bodies, alongside the suicided corpse of a captured SS prison guard, a bloated tongue protruding from his lips. Although the context of the images in a women's fashion magazine may appear incongruous to present-day viewers, *Vogue* had often included war-related content and employed a number of female war correspondents, including Miller.[46] Even so, her disturbing images exceeded any that had been published before, indicating the degree to which the Nazi atrocities had become a topic of national debate. The starkly graphic layout of unembellished, uncropped images, which appear to be hastily printed and still bear the marks of telegraphic transmission, present the images as unmediated evidence from the camps. The essay's title, 'Believe It', urges the viewer to reconcile the surreal and nightmarish images as truthful documents.

On other occasions the Allied authorities strategically hid photographs from sight, lest their actions be challenged. Audiences in the U.S. did not view the on-the-ground devastation wreaked by atomic bombs dropped on Japan on 6 and 9 August 1945, despite the existence of photographic documentation. Yoshito Matsushige, a photojournalist in Hiroshima, had made a handful of pictures just hours after the blast; Yōsuke Yamahata, a photographer for the Japanese News and Information Bureau, made a more complete document of Nagasaki the day after.[47] His images are an apocalyptic vision of devastation: a barefoot figure – perhaps an elderly woman – crawls in bewilderment;

84 Yōsuke Yamahata, Nagasaki, Japan, 10 August 1945, silver gelatin print.

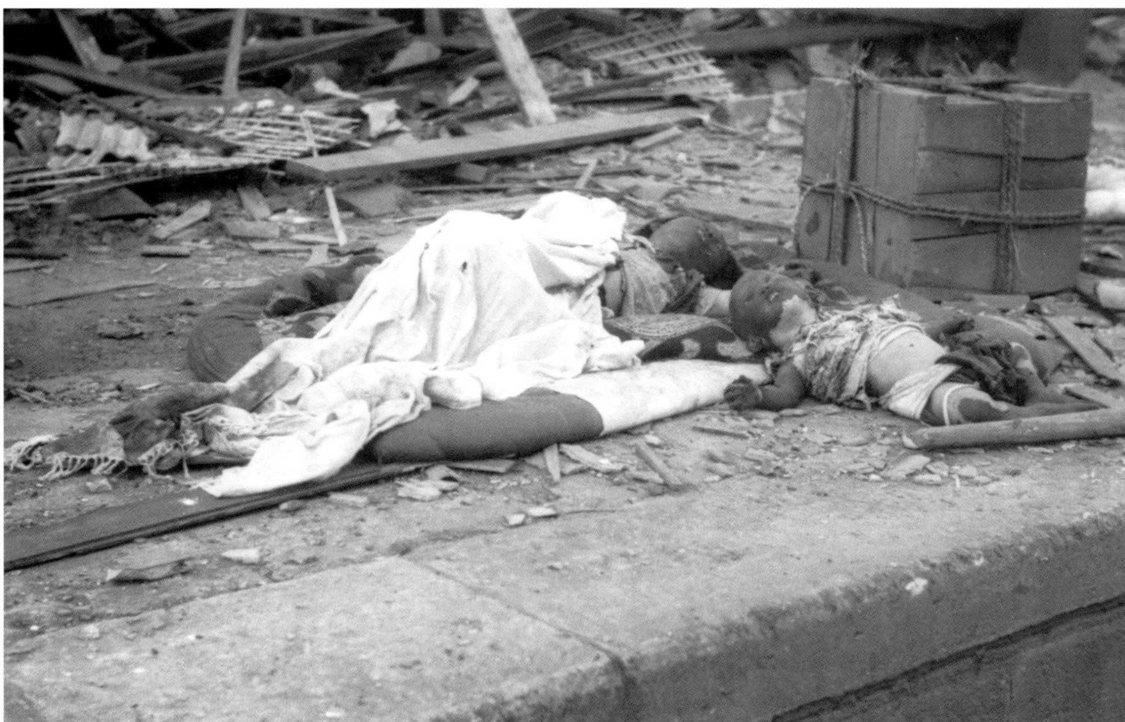

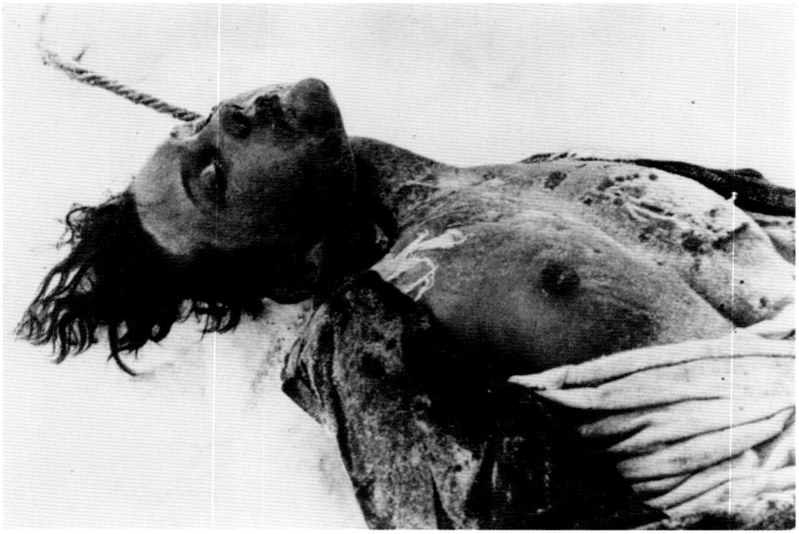

85 Sergei Strunnikov, *Tania*, 1941, silver gelatin print.

a mother spattered in blood attempts to breastfeed an injured baby; and charred and twisted bodies are strewn in the streets (illus. 84). Such potentially affecting images were anathema to U.S. wartime propaganda that styled the Japanese as an alien, if not subhuman, race. The authorities censored their publication, and it was not until 1952, when U.S. troops left Japan, that the images surfaced in the Japanese illustrated magazine *Asahi Gurafu* and subsequently in *Life* in the U.S.[48]

The exposure of repugnant images is thus predicated upon political expediency, and photographs of wartime sexual violence are no exception. Rape, assault and sexual torture have been – and continue to be – part and parcel of warfare, whether prohibited, tacitly approved or even actively encouraged.[49] Such violence is by no means restricted to acts by men against women: as the photographs of Abu Ghraib have shown, men may be the victims of abuse, and women the perpetrators. However, systematic sexual violence that aims to dominate, punish or even (in the case of the Rwandan Genocide) annihilate ethnic groups tends to be inflicted on women.

While such violence is a common occurrence in war, photographs are generally made public only to be paraded as proof of the enemy's barbarity. The Russian photographer Sergei Strunnikov made a distressing and

86 Ronald Haeberle, victims at My Lai Massacre, Vietnam, 1968, colour photograph.

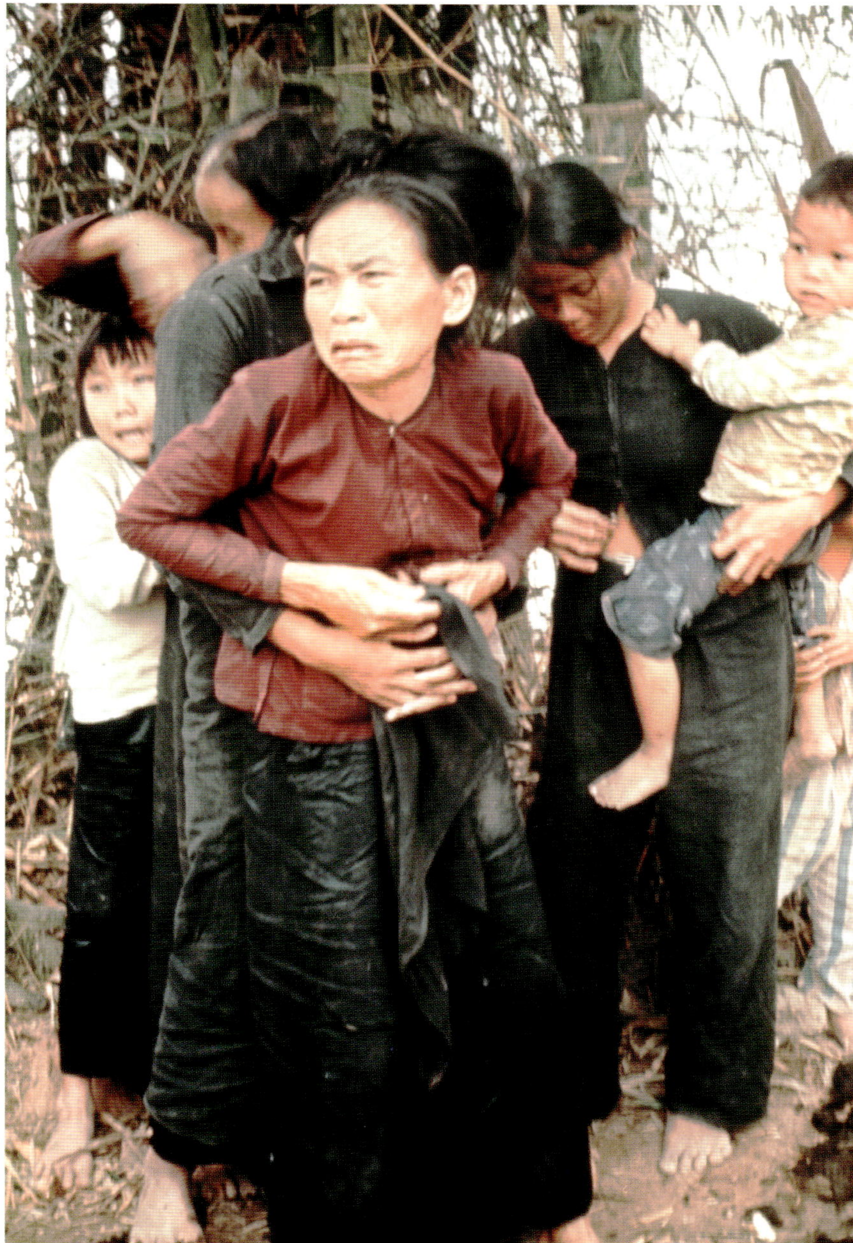

powerful image of the corpse of a young woman lying in the snow, a noose around her neck and her left breast mutilated (illus. 85). His image of 'Tania' (later identified as eighteen-year-old Soviet partisan Zoya Kosmodemyanskaya) was published in the Soviet newspaper *Pravda* in 1942, causing national outrage and transforming the young woman into a martyr and heroine. Strunnikov's image inspired a number of posters produced by the Soviet news agency TASS, including Viktor Deni's *Kill the Fascist Monster!* (1942), depicting a bulldog-like Nazi officer drooling over the teenager's corpse.[50] Subsequently, snapshots were allegedly found on the body of a dead German soldier, apparently showing the woman's last moments as she was hanged. Published in *Pravda* in 1943, the five German trophy snapshots were presented as proof of Nazi barbarity, and accompanying text urged Soviet citizens to harden their resolve to defeat their enemy.

By contrast, sexual assaults perpetrated by 'our boys' are highly unlikely to be acknowledged. Endemic rape and sexual violence in the Vietnam War was, and remains, a taboo subject. One of the most memorable images from the 1968 My Lai Massacre, made by military photographer Sgt Ron Haeberle in a private capacity, shows a group of women and children in anguish and confusion (illus. 86). As gender scholar Valerie Wieskamp has asserted, while the image is generally presented as (an admittedly distressing) document of civilians cowering before being machine-gunned to death, it in fact shows the anguish of a mother who has fought to keep her teenage daughter – who stands behind her buttoning up her blouse after being partially stripped – from being raped by U.S. soldiers.[51] Although initially reported in the 1969 *Life* magazine article, sexual violence at My Lai has been virtually expunged from American memory.[52] The gang rapes and torture reported in the 1969 Peers Inquiry did not lead to prosecutions, and subsequent retellings of the incident, such as the major PBS television documentary *My Lai* (2010), have avoided the subject altogether. Wieskamp makes the acerbic point that 'sexual violence is apparently more detrimental to the U.S.'s reputation than mass killing.'[53]

The use of photography to condemn wartime sexual violence presents conceptual challenges. Literal depictions – such as the image of the mutilated Soviet partisan 'Tania' – may unwittingly echo the violence already inflicted, or even be repurposed to serve a pornographic function. The Kurdish Iraqi

87 Seivan M. Salim, '27-year-old Delvin, who was captive for four months', from the series *Escaped*, 2015, C-type print.

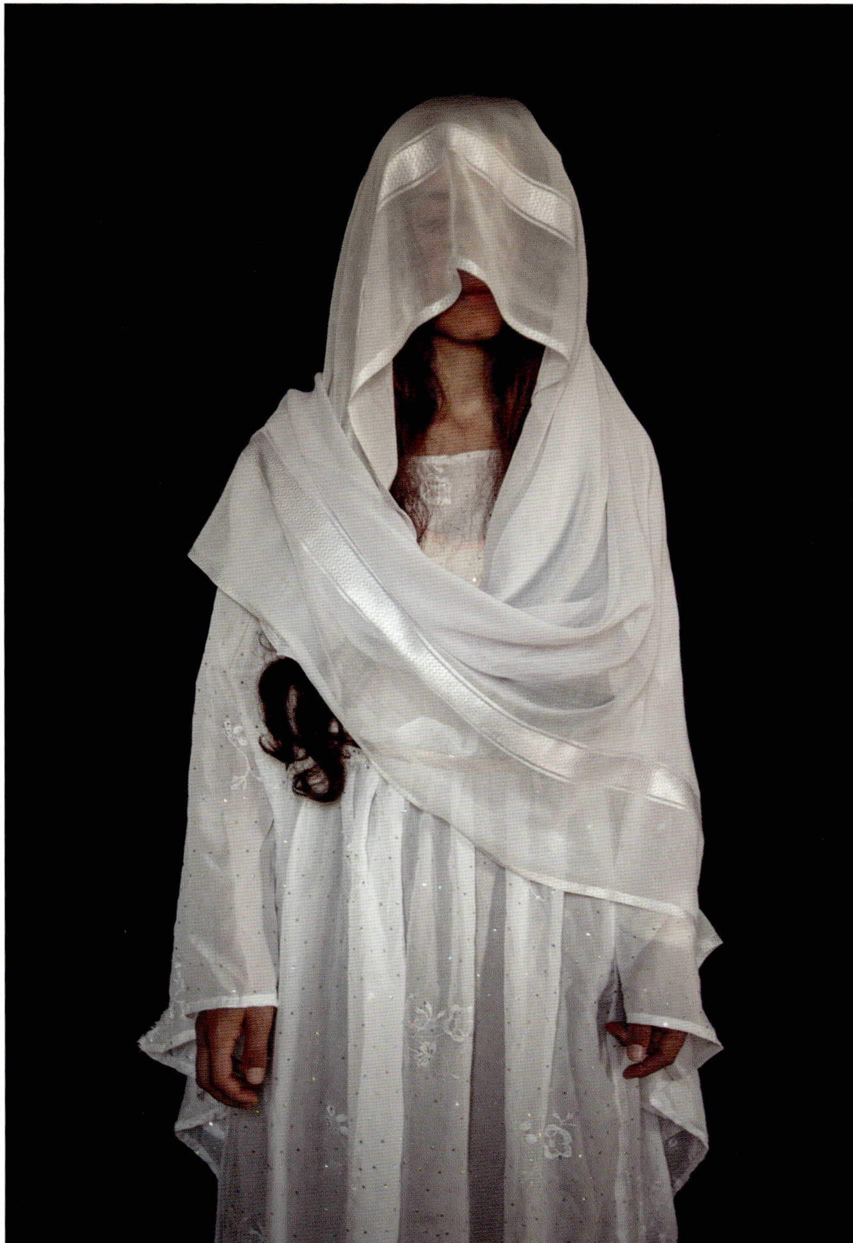

photographer Seivan M. Salim has been reflective of these concerns in *Escaped* (2015), a series of photographs addressing the recent sexual violence committed by ISIS militants against Yazidi women in the northern Iraqi city of Sinjar/Shingal. Tens of thousands of Yazidis were killed as part of an ethno-religious genocidal campaign, and an estimated 5,000 women kidnapped and sold into sexual slavery. Salim met some of the women who were able to break free.[54] Rejecting a photojournalistic approach, she chose to make portraits of the women wearing white clothing that evokes traditional Yazidi bridal garments, in contrast to the Muslim *abayas* (black robes) their captors forced them to wear (illus. 87). The veiled portraits form a muted counterpart to the women's graphic and harrowing testimonies. 'Delvin', a captive for four months, recalls,

> I was pregnant and I had other children with me. They were very cruel to us. Even though I was pregnant they would beat me and try to have sex with me. If I didn't accept to have sex with the men of the family, they would force me anyway. They raped me over and over again.

Shrouded and anonymized, the subjects of Salim's images are both spectral and corporeal, suggestive of conflicting desires to speak out while maintaining modesty in a society where femininity is predicated on purity. Salim's approach is consistent with the ethos of 'aftermath photography', whereby artists and conceptual photographers seek to avoid explicit depictions of violence, preferring more oblique strategies that seek to dignify the victims and leave space for the viewer's reflection. The aftermath approach is discussed in more depth in the next chapter.

Calls to Action: Truth, Justice and Reform

Clearly, photography may play an important role in excavating suppressed histories of violence during armed conflict. For some, however, it is not enough that photographs should simply expose atrocities and war crimes. Activists, pacifists, reformers and lawyers have deployed photography in their campaigns to call for systemic change or legal redress to prevent such violence happening again. The British campaigner Emily Hobhouse witnessed at first hand the concentration camps run by British forces in South Africa during the Second Boer War (1899–1902). According to Hobhouse, Afrikaner women and children were being held in dire conditions, subject to malnourishment and disease, and dying in their thousands. As part of an ongoing campaign, Hobhouse acquired photographs made in the camps in order to lend impact and authority to her reports.[55]

Such evidence could not be taken for granted, however. One of the images collected by Hobhouse depicted a naked girl named Lizzie van Zyl in a state of extreme emaciation (illus. 88). The obvious pathos of a dying child helpless on a bed, her eyes meeting the gaze of the onlooker, was powerful in instigating debate in society and politics in Britain.[56] The influential British author Arthur Conan Doyle, along with

88 Mr de Klerk (attrib.), Lizzie van Zyl, Bloemfontein concentration camp, Orange Free State Republic, c. 1902, silver gelatin print.

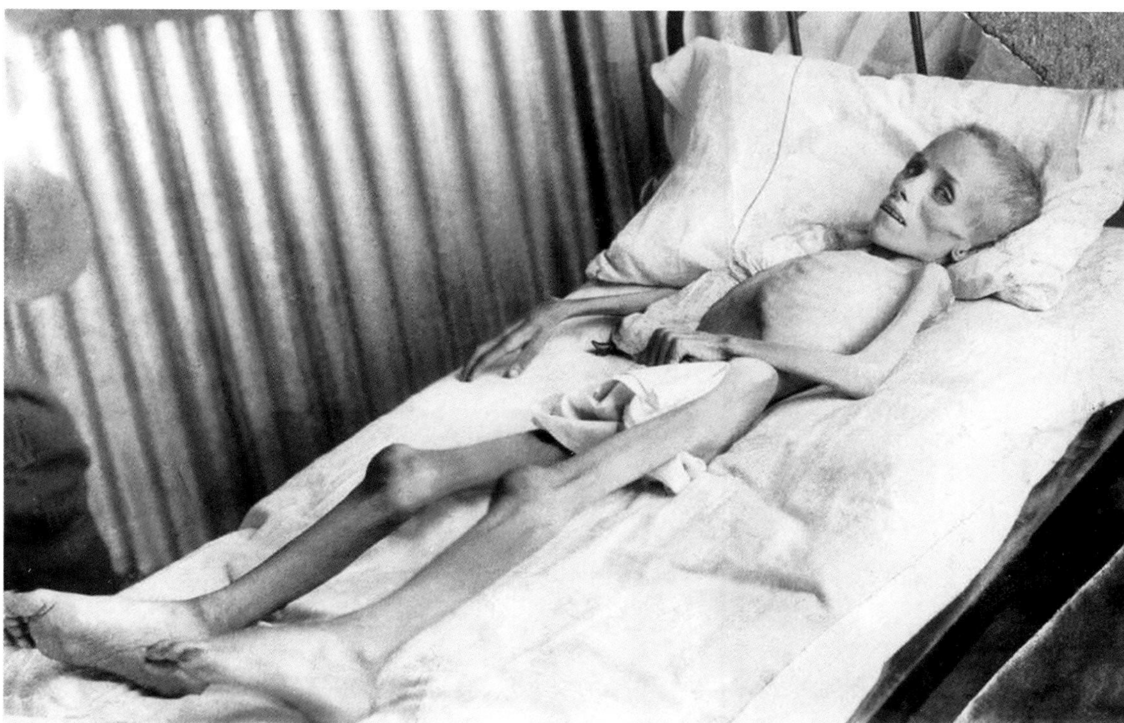

„De oorlog bekomt my als een kuur" (Hindenburg).

„War agrees with me like a stay at a health resort." (Hindenburg.)

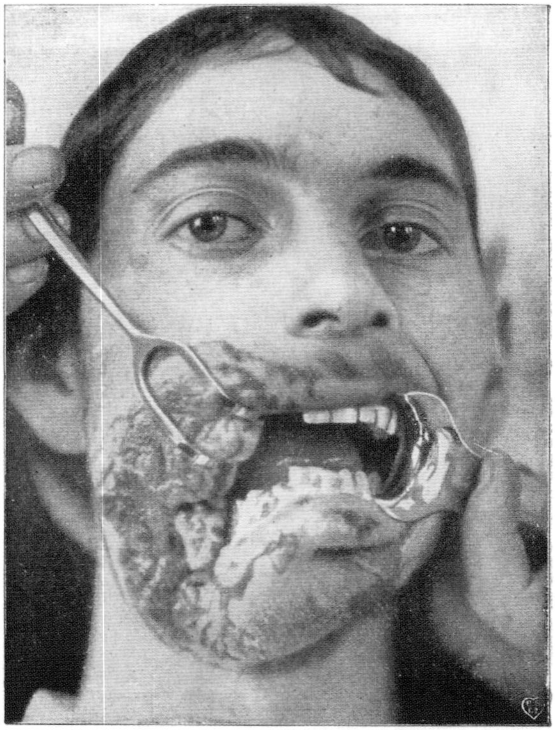

„Der Krieg bekommt mir wie eine Badekur." (Hindenburg.)

'La guerre est pour moi un traitement d'eaux minérales.' (Hindenburg.)

89 Ernst Friedrich, from the publication *Krieg dem Kriege!* (War Against War!), Berlin, 1930.

the pro-government British press, claimed that the child had already suffered from malnourishment before entering the camp, and was in fact a victim of maternal neglect. In Hobhouse's 1902 publication, *The Brunt of War and Where it Fell*, she counters such claims by tracking down the photographer – an inmate known to us only as Mr de Klerk – to confirm he had made the photograph two months after Lizzie van Zyl had entered Bloemfontein camp. In a footnote to her text, Hobhouse admonishes her publisher, who considered the image 'too painful for reproduction'. In a succinct but persuasive rhetoric, Hobhouse questions 'how far it is right to shrink from a typical representation, however distressing, of suffering, which others have to endure, and which has been brought about by a sequence of events for which we are partly responsible': an ethical question which remains pertinent.[57] Despite provoking the ire of the British authorities, Hobhouse's campaign was instrumental in initiating an enquiry into the camps, and contributed to improvement of conditions.

One of the most dogged campaigners to use photographs in the quest to end suffering was the German pacifist Ernst Friedrich. His self-published *Krieg dem Kriege!* (War Against War!, Berlin, 1924), remains one of the most important photobooks to consider armed conflict. Friedrich fervently believed that if people could see the real horrors experienced by soldiers in the First World War, further conflicts would be prevented. Mechanized warfare meant that millions of young men had been maimed or killed, yet images of dead or injured soldiers had generally been suppressed in the mainstream media. Society was not ready to accept the realities of young men returning from the front with missing limbs, genitals or jaws. Veterans with obvious physical disabilities or facial disfigurements might find themselves the objects of pity, and were in some instances shunned from post-war parades or memorial ceremonies, their physical imperfections a disturbing reminder of the shattered assumption of the desirable masculinity of soldiering.[58]

Drawing largely on images from German medical and military archives, Friedrich compiled *Krieg dem Kriege!*, a cabinet of horrors, the centrepiece of which is a section on facial injuries. Page after page depicts men with parts of their faces shot away. Jaws and throats are blown apart to reveal the soft interior; tongues lie limp; and eyes and noses are missing. The images of disfigured men – still alive despite their injuries and aware of the photographer's gaze – are a powerful antidote to war mythology of the 'beautiful dead', typified by the aesthetics of wholeness and symmetry perpetuated in official memorials.[59] Captions in four languages, some drawn with heavy irony from pro-war propaganda, leave the reader in no doubt as to the book's mission.[60] Field Marshal Hindenburg's declaration that 'War agrees with me like a stay at a health resort', for instance, is paired with an image of a man with a grotesquely shattered jaw (illus. 89). Friedrich subsequently reprinted the book in various formats to raise funds to support his lifelong crusade to end war. He also exhibited photographs in his

Internationales Anti-Kriegsmuseum, the world's first anti-war museum, which he founded in 1923.[61] Unsurprisingly, his activities brought him into conflict with the authorities, and he was forced to leave Germany in 1933.[62] His continued persecution by German authorities, including a death sentence *in absentia*, indicates the significance of his photographic activism.

Civilians' photographs became the unexpected catalyst for one of the most resourceful and imaginative campaigns for truth and justice in the twentieth century. In 1976 a group of Argentine mothers banded together to seek information about their missing children who had, it transpired, been abducted, illegally detained and in some instances killed by the secret police of the far-right military junta. To assist in their searches, the women carried pictures of their missing loved ones to serve as aids to identification. When the authorities dismissed their pleas, the women began to hold up family photographs and ID cards during their weekly demonstrations in Plaza de Mayo, outside the seat of government in Buenos Aires. Although vilified by politicians as 'mad women', the Madres de Plaza de Mayo (Mothers of May Square) continued utilizing photographs of the disappeared on an increasing scale, featuring them on banners and placards throughout the period of state terrorism and afterwards during the post-dictatorship era.[63]

Visibility was key to the campaign, and the women collaborated with sympathetic press photographers to ensure their protests reached wider audiences.[64] An image by Argentine photojournalist Eduardo Longoni depicts the spectacle of women demonstrating at Plaza de Mayo (illus. 90). One woman, her eyes closed, holds up a snapshot of a young woman seated playfully on a desk, legs swinging. The young woman is Graciela Mellibovsky, a 26-year-old government statistician abducted in 1976, and the woman holding the snapshot is her mother, Matilde.[65] The photogenic spectacle of such protests, and the emotional power of brandished family snapshots, soon received attention in the international press, contributing to growing pressure for a truth commission. The National Commission on the Disappearance of Persons (CONADEP) was duly launched following the collapse of the junta in 1983.

Photographs have remained important in the ongoing struggles for truth and justice in Argentina.[66] In the 1990s the Madres de Plaza de Mayo Línea Fundadora held outdoor demonstrations in public parks in Buenos

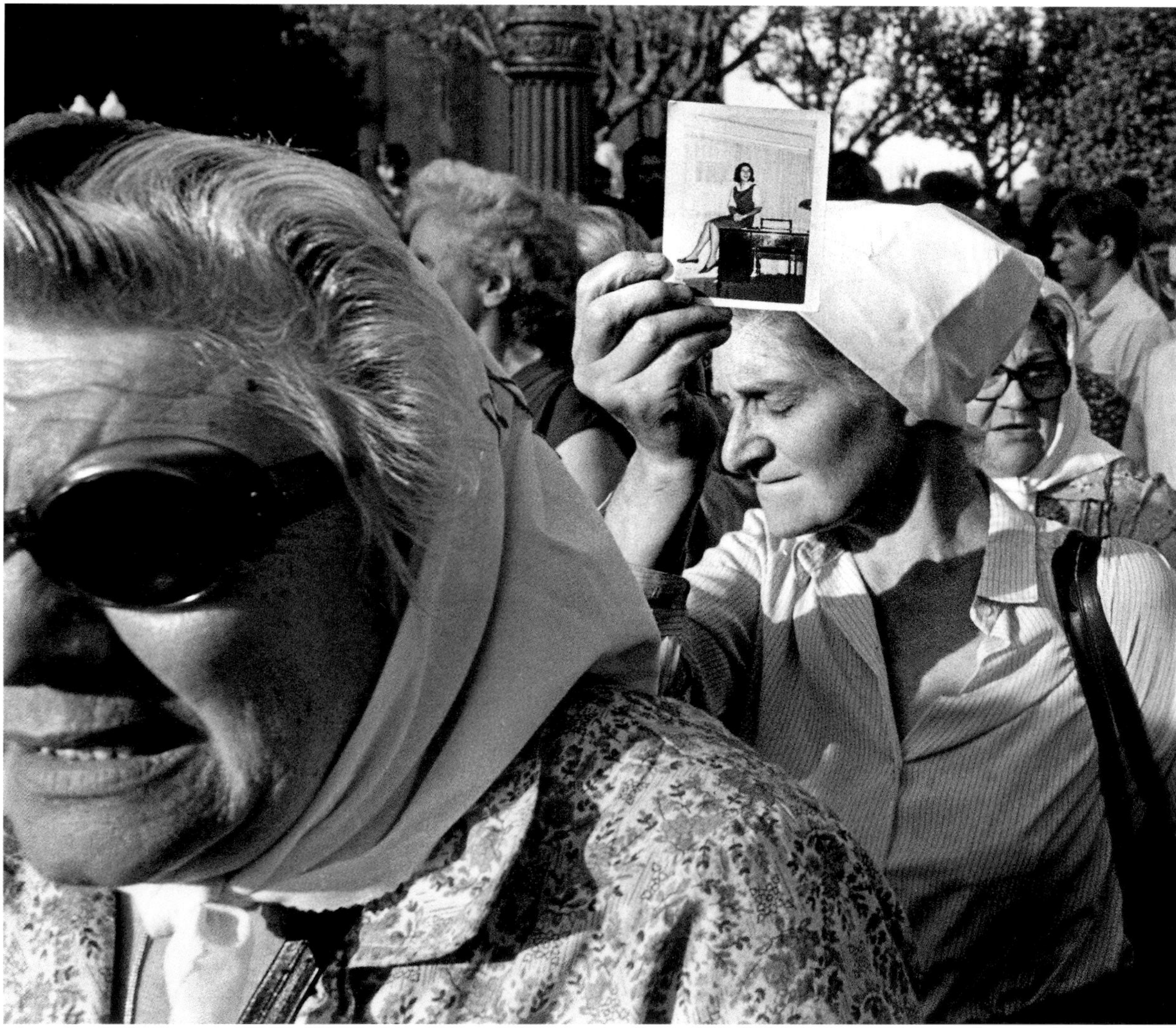

90 Eduardo Longoni, Madres de Plaza de Mayo marching in Buenos Aires, 1981, silver gelatin print.

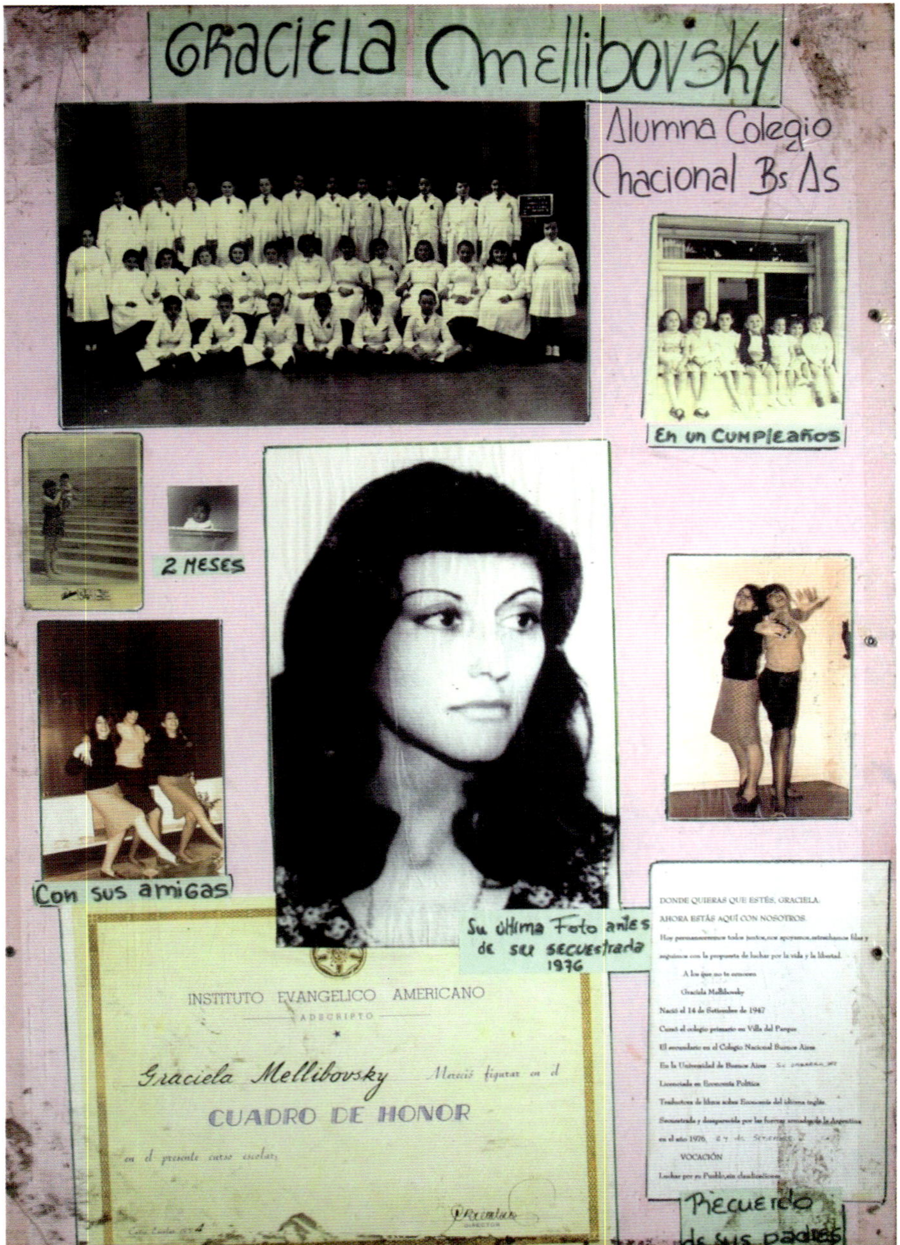
GRACIELA MELLIBOVSKY
Alumna Colegio Nacional Bs As
En un Cumpleaños
2 MESES
Con sus amigas
Su última Foto antes de ser secuestrada 1976
INSTITUTO EVANGELICO AMERICANO
ADSCRIPTO
Graciela Mellibovsky Mereció figurar en el
CUADRO DE HONOR
en el presente curso escolar
DONDE QUIERAS QUE ESTÉS, GRACIELA
AHORA ESTÁS AQUÍ CON NOSOTROS
Hoy permanecemos todos juntos, nos apoyamos, estrechamos filas y seguimos con la propuesta de luchar por la vida y la libertad.
A los que no te conocen
Graciela Mellibovsky
Nació el 14 de Setiembre de 1947
Cursó el colegio primario en Villa del Parque
El secundario en el Colegio Nacional Buenos Aires
En la Universidad de Buenos Aires
Licenciada en Economía Política
Traductora de libros sobre Economía del idioma inglés
Secuestrada y desaparecida por las fuerzas armadas de la Argentina
en el año 1976,
VOCACIÓN
Luchar por su Pueblo, sin claudicaciones
Recuerdo de sus padres

Aires where relatives could present pictures of their missing loved ones and speak to passers-by.[67] Matilde Mellibovsky created a placard featuring family photographs of her daughter Graciela, including a baby portrait captioned 'two months', a snap of a toddler at a birthday party, and a clumsily composed shot of three laughing teenagers (illus. 91). At the centre is a portrait of Graciela appended with the haunting caption: 'Her last photo before being abducted'. Mobilizing the language of the family album, the placard invites spectators to imagine themselves in the guise of family members, and empathize with Matilde's loss. The images are about as far from conventional war photography as could be imagined, yet their very ordinariness is key to presenting Graciela as a normal person whose life was linked to the social structures of family, school and work, a life photographically punctuated by the conformity of birthday parties and trips to the seaside. Counteracting the state's accusations of 'subversion', this humble placard quietly but forcefully attests to the victim's humanity and moral courage in opposing the regime.[68]

The assumed evidential capacity of photography – however problematic in reality – has given it an important role in post-war judicial processes. In Guatemala, for instance, the National Police Historical Archive (AHPN) has made ID photography by state perpetrators available to researchers and legal investigators. Alongside other evidence such as dossiers on victims and exhumed bones, these photographs are helping to identify some of the 200,000 estimated persons that were disappeared during the decades of genocide, and to support the slow and contested process of prosecutions.[69]

On other occasions, photographs have been made specifically to perform work in law courts. Photographic prints of Ravensbrueck concentration camp survivor Jadwiga Dzido, a member of the Polish underground resistance, were among those presented to the jury at the Medical Trial at Nuremberg in 1946, as evidence of Nazi medical experiments carried out on prisoners (illus. 92). Clad in a white sheet, Dzido is photographed from various angles, displaying her scarred and damaged leg, which was infected with dirt and slivers of metal to simulate battle wounds prior to treatment with experimental drugs. The easy reproducibility and dissemination of photographic images, coupled with their assumed veracity and objectivity, meant they were – and remain – ideally suited to function as legal evidence, passed around court chambers and appended to reports.

91 Matilde Mellibovsky, *Graciela Mellibovsky*, c. 1990, placard with photocopied photographs and annotations.

In 2014 images showing victims of Bashar al-Assad's regime in Syria were scrutinized in the United States Capitol in Washington, DC. The 'Caesar Photographs', titled after the codename of the anonymous Syrian military forensic photographer who leaked them, claimed to provide evidence of starvation, mistreatment and torture on a vast scale: perhaps as many as 11,000 deaths between 2011 and 2013. The images are strictly functional, with the hand of the photographer sometimes visible in the corner of the frame holding a card on which is scrawled reference numbers pertaining to the victim's identity. The images were presented as part of the briefing 'Assad's Killing Machine Exposed: Implications for U.S. Policy', held by the House Foreign Affairs Committee in 2014.[70] A press image of the briefing (illus. 93) shows Syrian Army defector 'Caesar', his identity concealed by a blue hooded jacket, listening to the Committee through an interpreter. Large prints of the victims, their features digitally blurred or masked, are displayed for the briefing. The photographs were instrumental in the passing of the Caesar Syria Civilian Protection Act of 2017, which pledged to 'halt the wholesale slaughter of the Syrian people, encourage a negotiated

92 Unidentified photographer, Jadwiga Dzido displaying injuries caused by Nazi medical experiments, court documents submitted at Nuremberg, 17 December 1946.

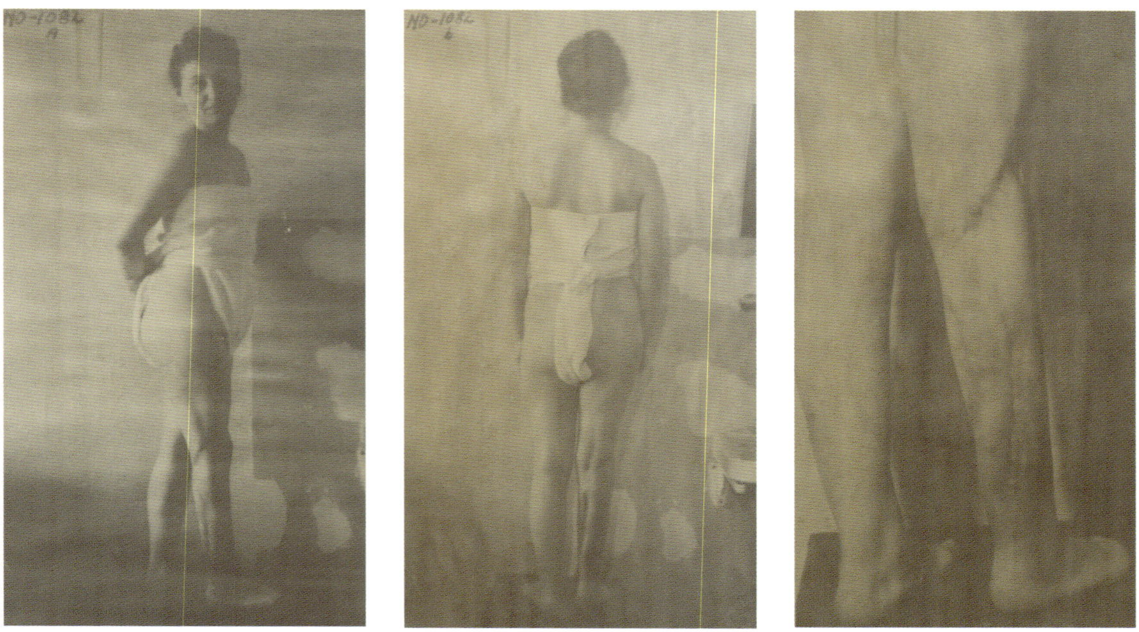

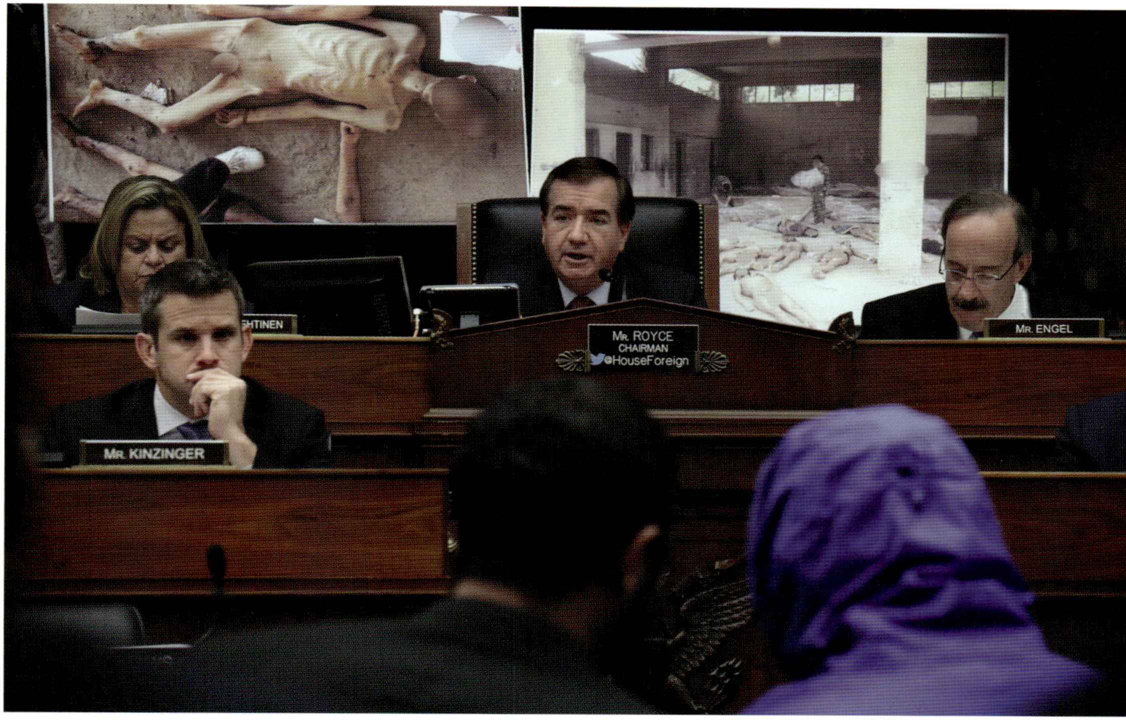

93 Alex Wong, former Syrian Army photographer 'Caeser' attends a hearing at U.S. House of Foreign Affairs Committee, 2014, colour digital photograph.

political settlement, and hold Syrian human rights abusers accountable for their crimes'.[71] While it remains to be seen if the Act results in a real difference for those most affected, it demonstrates a positive step towards action.

As demonstrated by the ongoing legal and humanitarian processes in Argentina, Guatemala and Syria – as well as many other regions – the effects of war may reverberate long after the violent event has taken place. The ways in which photography has responded to the legacies of war is explored next.

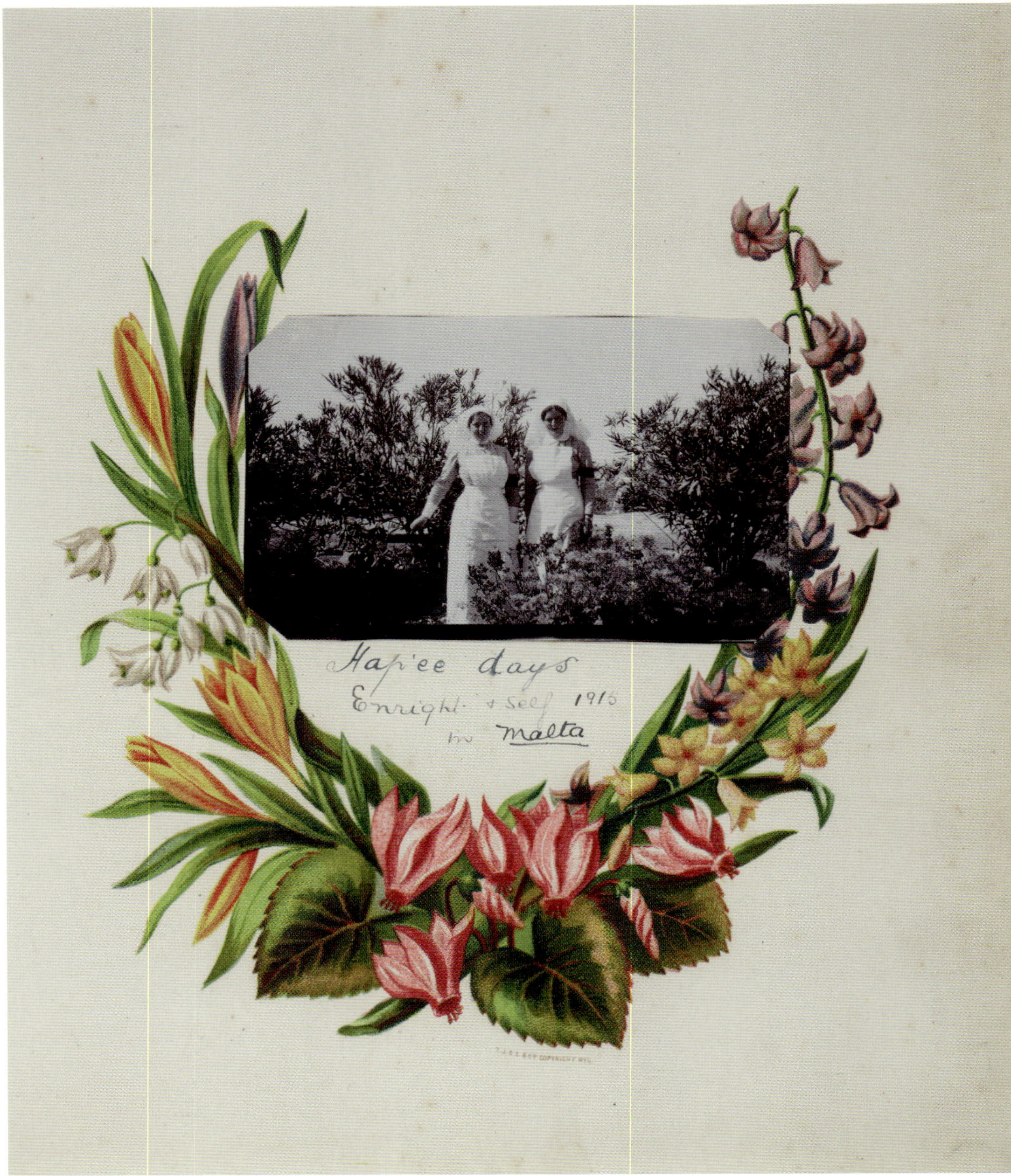
Hap'ee days
Enright + Self 1915
in Malta

five

Legacies

War shapes lives. For many people, the end of fighting does not bring a return to the pre-war past, but signals a new chapter. There may be consequences both for individuals and for societies, sometimes felt immediately, sometimes emerging years later. Photography has a powerful role to play in documenting, remembering and disseminating the legacies of war, as well as in transitions to peace or, at least, to relative security. The photographs discussed in this chapter draw attention to the material remains of war as well as its intangible after-effects, and connect with ongoing processes of personal mourning, national remembrance and post-war rehabilitation.

Portraiture, Veterans and Survivors

For some veterans, wartime military service may be a defining experience that still shapes their identity, powerfully expressed through photographic portraits. Some of the earliest depict officers that served in the Napoleonic Wars. The former soldiers, now elderly men, appear proudly attired in ceremonial uniform, decorated with medals (illus. 95). Grenadier Burg of the 23rd Regiment of the Guard, who sports a splendid white moustache, stands to attention with bearskin and rapier, his white trousers stretched over his now-portly middle. Photography had not been invented at the time of the Napoleonic Wars, and these anachronistic portraits serve to both belatedly visualize military service and to form a tangible link to the pre-photographic past.

94 Sister Sara 'Josie' Kane, 'Hap'ee Days: Enright and Self in Malta, 1915', silver gelatin print in photo album.

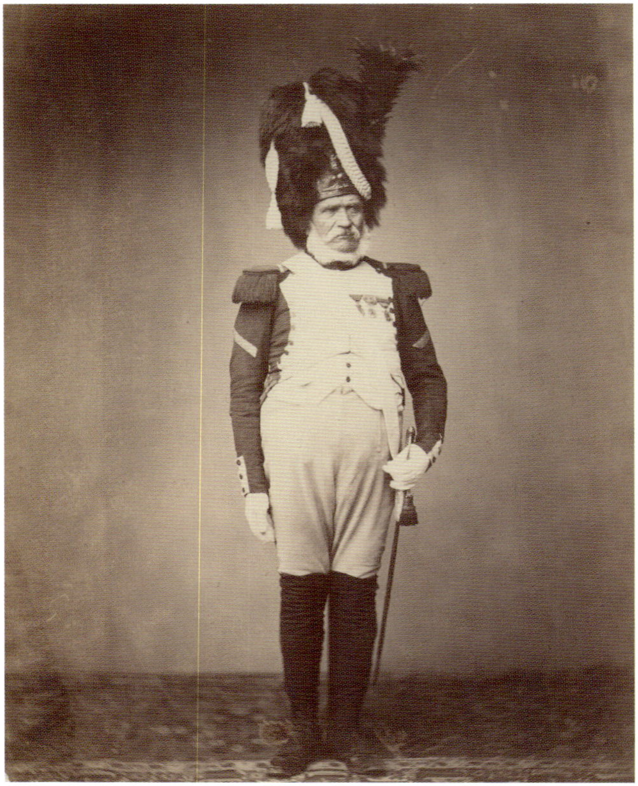

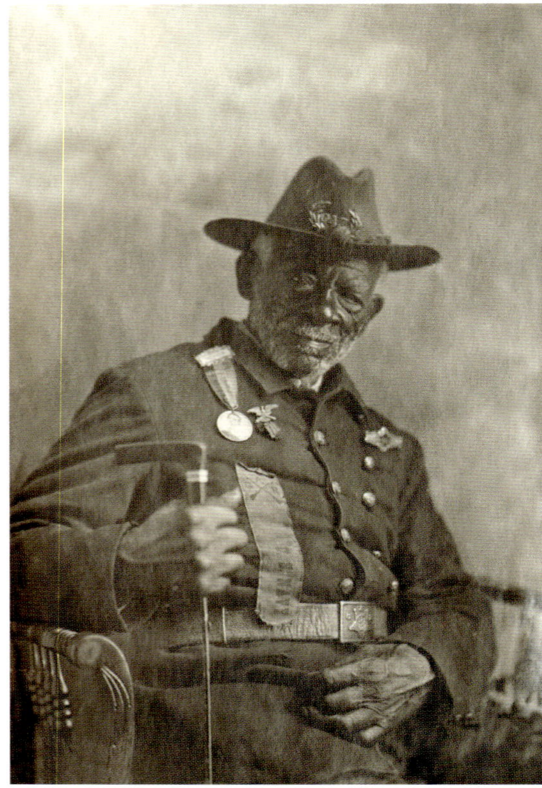

A contrasting portrait made in 1925 depicts Smart Chisholm, a veteran of the American Civil War (illus. 96).[1] Made by Elise Harleston, an African American photographer working in Charleston, South Carolina, this image is more intimate and humble than the formal portrait of the Napoleonic Grenadier. Chisholm's posture is relaxed, almost slumped. A half-smile plays on his lips, and his eyes gaze out in an ambiguous expression of reverie. Harleston's shallow depth of field softens the foreground and background, allowing the viewer to dwell on telling details: Chisholm's grizzled beard, his wrinkled hand resting on his lap and the frayed Grand Army of the Republic ribbon, roughly tacked to his chest. As a Union veteran in a region that still venerated the Confederacy and endorsed racism, the portrait can be seen as an act of political conviction as much as a personal record of service.[2]

95 Grenadier Burg of the 23rd Regiment of the Guard of 1815, c. 1860, albumen sepia print.

96 Elise Harleston, *A Colored Grand Army Man*, 1925, silver gelatin print.

Although divergent in subject and viewpoint, both of these portraits mark the passage of time and the transformation of young warriors into elderly veterans. Soldiers killed in action, by contrast, remain frozen in their youth, sometimes leaving their photographs to be treasured by those left behind. A tintype made around the time of the American Civil War depicts a young girl in a plaid dress with her hands clasped protectively around a cased photograph, probably another tintype or an ambrotype (illus. 97). The picture-in-a-picture was a common device used by nineteenth-century portraitists to indicate remembrance of an ancestor or a deceased loved one. In this instance, the picture-in-a-picture depicts a cavalryman with a sword

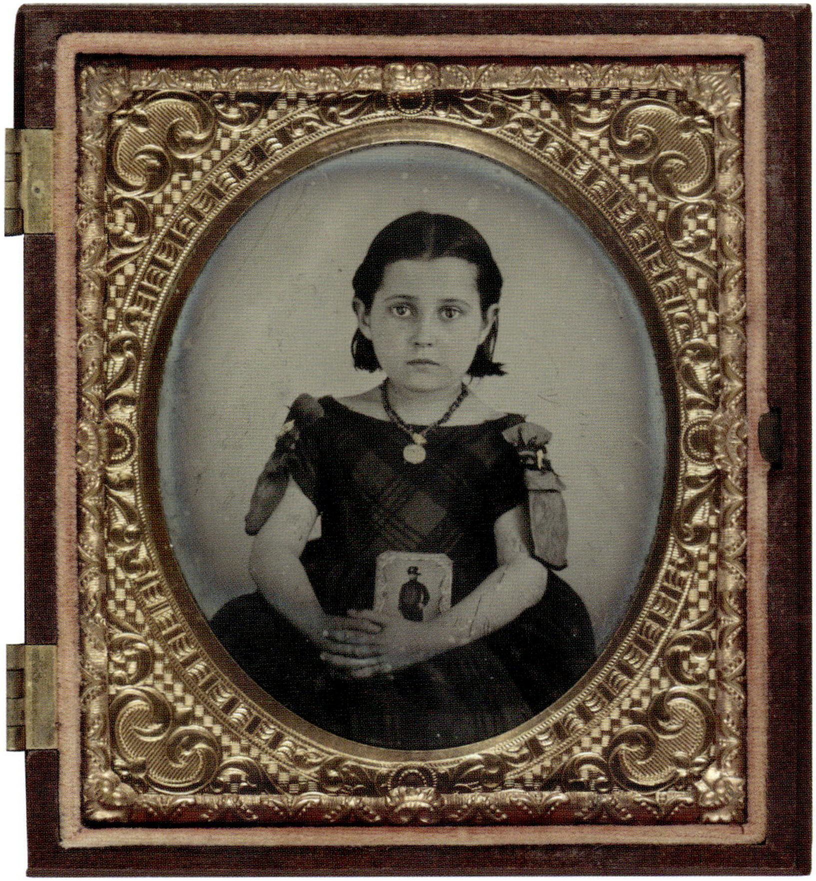

97 Unidentified photographer, girl in mourning dress, c. 1861–70, sixth-plate tintype, hand-coloured, in case.

and the Hardee hat of the Union Army, presumably the father of the girl in the portrait. Whether or not he returned from fighting is not known, but the girl's dress, necklace and mourning ribbons suggest he died in the conflict.[3] Tintypes and ambrotypes, which flourished during the conflict, were unique positives made on iron or glass plates respectively, and as no negative existed, they were irreplaceable, making the likeness of the absent soldier even more precious. The object demonstrates the peculiar potency of photographs to stand proxy for those no longer with us, offering comfort to the mourner as well as an aid to memory.

The trope of the picture-in-a-picture remains resonant. The Iranian photographer Newsha Tavakolian uses a similar approach in her series *Mothers of Martyrs* (2006), depicting elderly Iranian women holding up cherished portraits of sons killed in the Iran–Iraq War (1980–88) (illus. 98).

98 Newsha Tavakolian, from the series *Mothers of Martyrs*, 2006, C-type print.

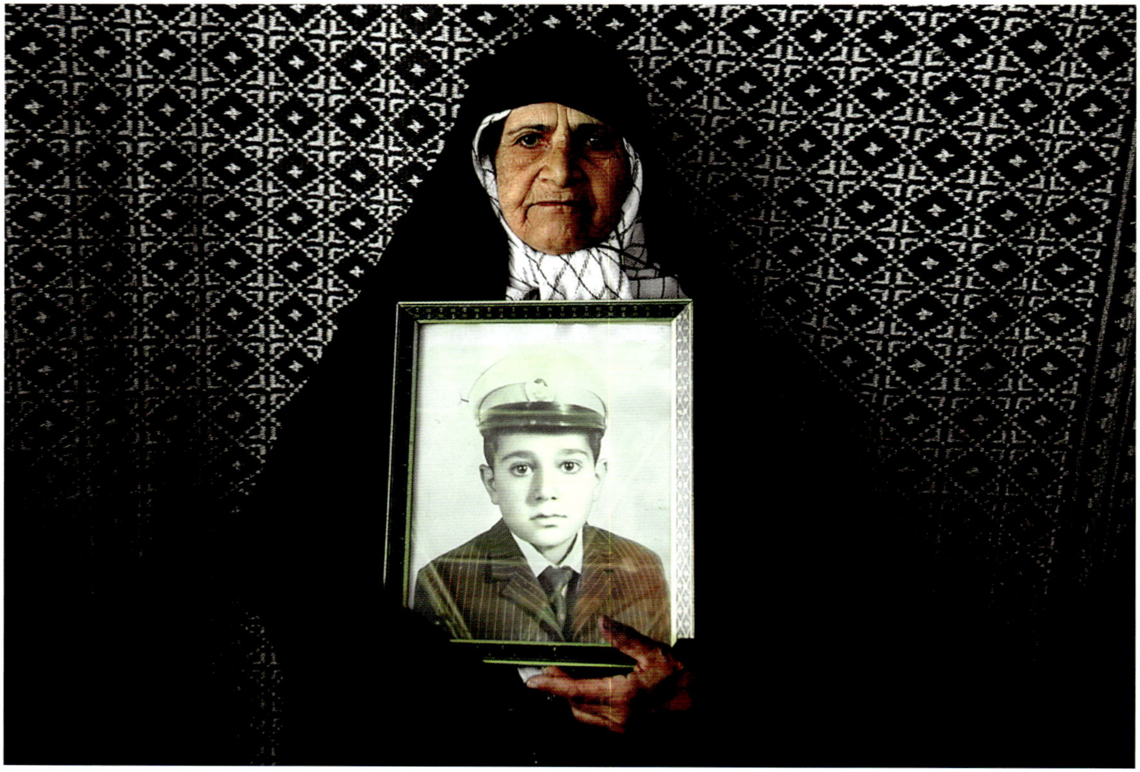

Their gestures recall the protests of the Argentine mothers who held up pictures of their children who had been 'disappeared' by the military dictatorship of 1976–83 (see illus. 90). The motive of the Iranian mothers, however, is ambiguous. Are their gestures demands for redress? Are the women resentful and angry to have lost their sons? Or are they proud to present the children who sacrificed their lives for their country, as the term 'martyrs' suggests? Tavakolian's series points to the complexity of motherhood in the context of militarism, as well as highlighting the poignant gulf between the ageing women and their forever-youthful sons.

Both the Civil War tintype and the framed portrait held by the Iranian mothers are manifestly physical objects, to be touched, passed round to family members and perhaps wept over. Photo albums are particularly rich memory-objects, combining photographs and captions in sequences that narrate momentous periods in the lives of their creators. Servicemen and women have often made such records, which can evoke memories of pride, camaraderie and adventure as well as more painful recollections. Sister Sara 'Josie' Kane, who served as a nurse in Malta in the First World War, compiled snapshots in a mass-produced album decorated with floral emblems (illus. 94).[4] Filled with records of her wartime work, Kane's album is a collision of masculine and feminine pictorial conventions. While the subject matter indicates the harsh realities of wartime nursing, such as an image of an amputee patient swathed in blankets, it is clear that Kane also experienced the war as a period of independence and utility, even a time she remembers as 'Hap'ee Days'.

Embellishment and creative intervention may help to increase the potency of photographs to serve as mementos of wartime experience. The curator Geoffrey Batchen argues that hybrid photographic objects, which combine the senses of touch as well as sight, are better able to conjure up the past.[5] One striking example of this is a double metal frame, made in the Second World War, in which two oval portraits are set: a man in a military uniform, and a civilian woman, identified only as Norma Lee, wearing a sweater and a string of beads around her neck (illus. 99). The two frames are separate but linked, suggesting an attachment between the two people despite, presumably, the vast distance between them. It is not immediately apparent that the diamond-shaped frames are in fact made from bullets, probably

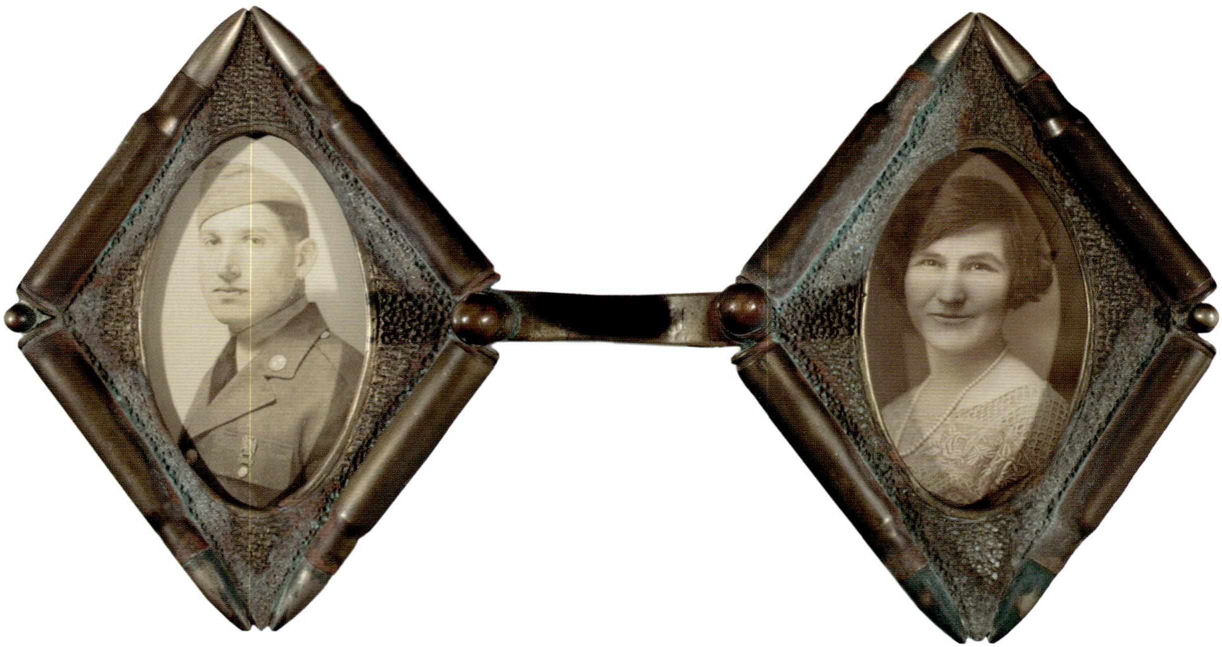

relics of war collected by the soldier. The materiality of the object lends it a greater immediacy, able to transport the viewer/holder into the military realm where it was fashioned. There is a great poignancy in this humble object, the care lavished on it, and the desire to make something of beauty and love out of the most unexpected materials. It is perhaps all the more poignant that this little object was subsequently cast adrift, retrieved by Batchen from a junk shop in New Mexico.[6]

99 Unidentified photgrapher, a soldier and 'Norma Lee', c. 1942, photographs in frame made from bullets.

Contemporary artists have creatively embellished photographs for strategic purposes. The Apsa'alooke artist Wendy Red Star has considered the role of her ancestral Crow tribe in resisting the Native American genocide during the Indian Wars of the 1860s to 1890s. Photographic depictions of the conflict were invariably made from the U.S. viewpoint, and depicted Native Americans as exotic, primitive or barbaric. In the face of this archival lack, Red Star has appropriated studio portraits of Crow leaders made by the American photographer Charles Milton Bell, on the occasion of the Crow visit to Washington, DC, to negotiate peace in 1880.[7] Transformed from *cartes de visite* into imposing large prints, and embellished with Red Star's

100 Wendy Red Star, *Déaxitchish/Pretty Eagle* from the series *1880 Crow Peace Delegation*, 2014, inkjet print and red ink on paper.

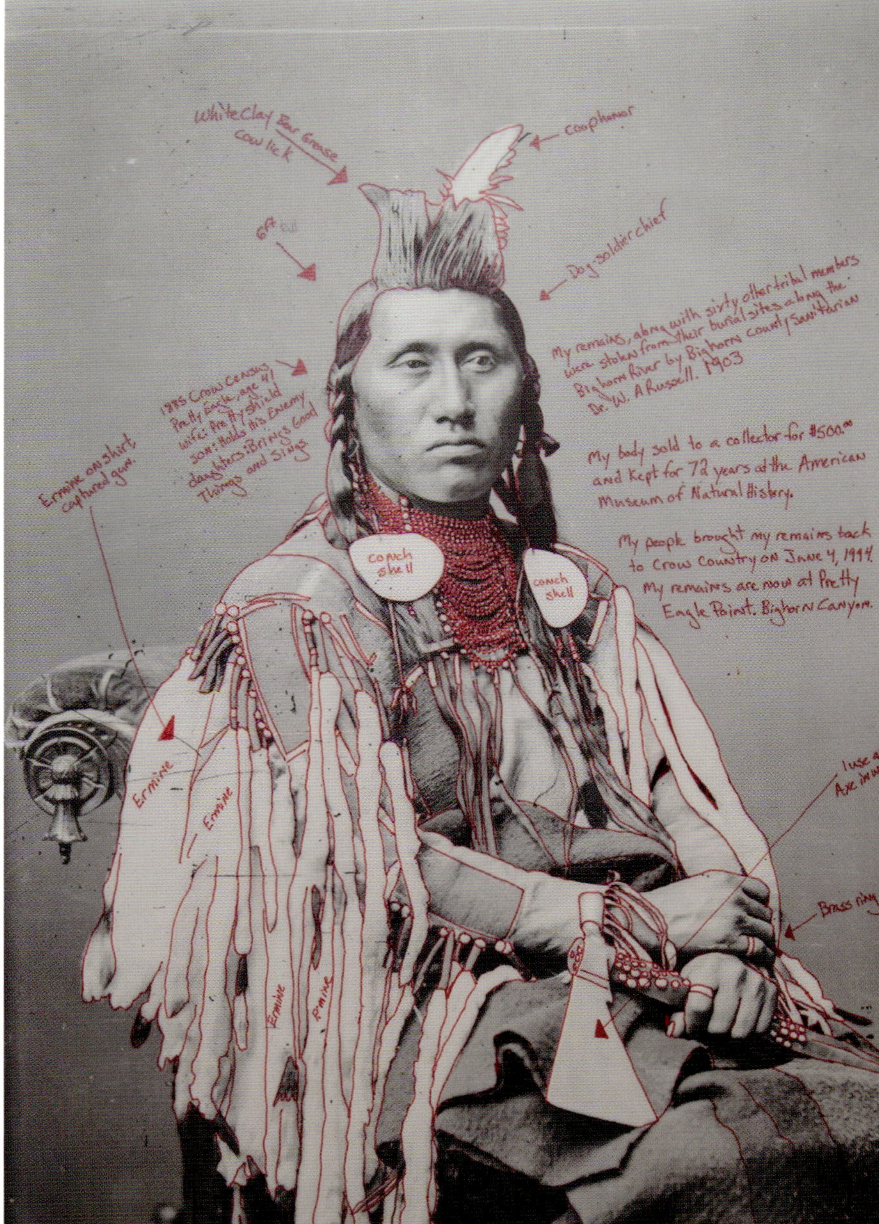
White Clay
Bear Grease
Cow lick
Coup honor
6 ft tall
Dog soldier chief
1885 Crow Census
Pretty Eagle, age 41
wife: Pretty Shield
son: Holds His Enemy
daughters: Brings Good Things and Sings
Ermine on shirt
captured gun.
My remains, along with sixty other tribal members were stolen from their burial sites along the Bighorn River by Bighorn county Sanitarian Dr. W. A Russell. 1903
My body sold to a collector for $500.00 and kept for 72 years at the American Museum of Natural History.
My people brought my remains back to Crow Country on June 4, 1994. My remains are now at Pretty Eagle Point. Bighorn Canyon.
conch shell
conch shell
Ermine
Ermine
Ermine
Ermine
I use an Axe in war
Brass ring

idiosyncratic red ink annotations, the portraits assert the individuality of the sitters and their agency as political negotiators who sought to end conflict (illus. 100). Reactivated over a century later, the images remind the viewer that the events of long-concluded wars continue to reverberate in the present, and that histories can be re-told.

Collective Memory, Commemoration and War Tourism

Memory is not, of course, restricted to individual acts, nor is it always a private activity. The extraordinary and traumatic events of wartime often give rise to the collective desire to commemorate, retrace and even relive what happened. Photographs have been important to communal acts of remembrance, particularly in the wake of national traumas such as genocide and state terrorism, which tend to be characterized by official denials and climates of impunity. As discussed in the last chapter, legal processes to prosecute the perpetrators are one course of action. The first step towards reconciliation, however, may simply be public acknowledgment of what happened. It can help to collect and display photographs within a shared space for reflection, mourning and comprehension. Many societies have constructed archives, databases and exhibitions, both physical and virtual, to recover and preserve collective memory, such as the Argentine project Memoria Abierta (Open Memory), begun in 1999.[8] Memory museums – such as the Tuol Sleng Museum of Genocide in Cambodia, discussed in the previous chapter – are among the many physical sites to offer permanent displays of photographs of victims and scenes of atrocities.[9] Galerija 11/07/95 in Sarajevo is dedicated to the memory of the massacre of Bosnian Muslims by Serbs in the small town of Srebrenica in 1995, during the Bosnian War (1992–5).[10] Black-and-white photographs are hung on the grey walls and suspended from ceilings, forming a sombre and immersive environment for the visitor (illus. 101). Although such projects run the risk of repeating the trauma of the past or even perpetuating rifts between communities, they may also provide a 'safe' space to reflect upon life-altering violence and honour those who are gone. For visitors from outside the community, memory museums

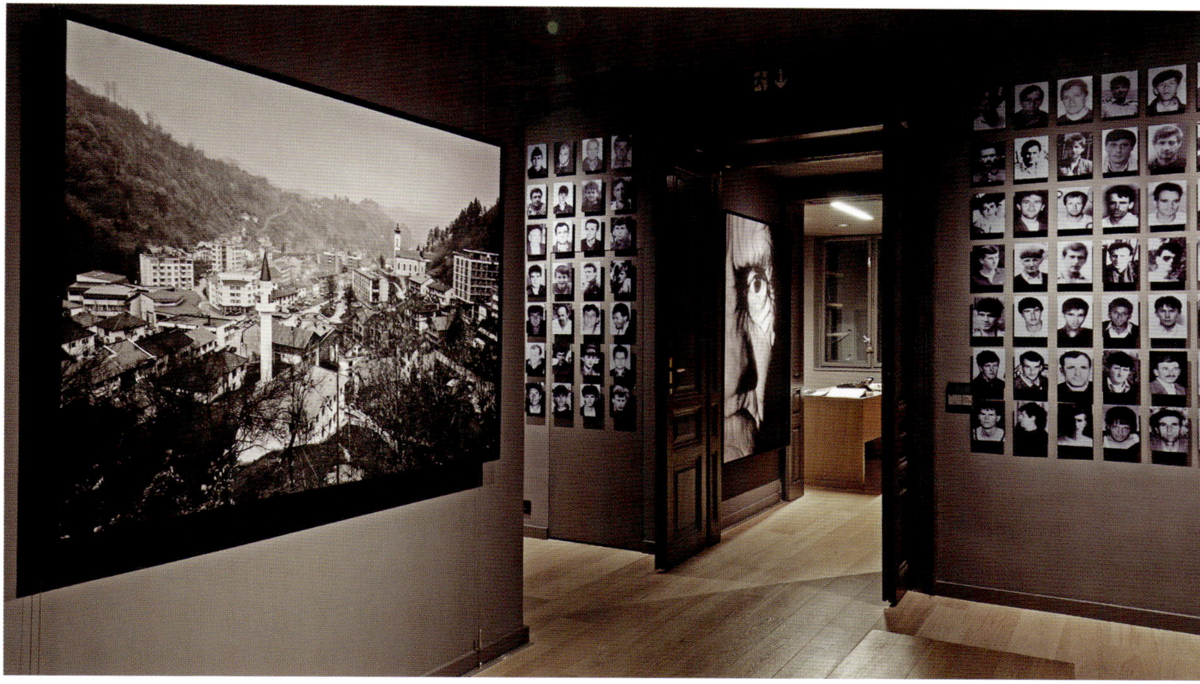

101 Images of victims of the Srebrenica massacre, displayed in Galerija 11/07/95, Sarajevo, Bosnia and Herzegovina.

offer the possibility of education and – perhaps – the desire to prevent the repetition of such occurrences.

Anachronistic engagement with historical conflicts, however, is fraught with problems, and contemporary artists have used photography to observe and critique the ways in which the past is replayed in the present. Battle re-enactment straddles the fields of education and entertainment. 'Living history' groups in Europe and the U.S. offer a simulated experience of danger, drama and heroism, where participants dress in period costume and contribute to communal live performances in which they act out roles, such as soldier, nurse or spy. The potential absurdity of such activities has been the target of photographers quick to spot the dissonance between the historical fiction and present-day reality. A portrait from Jonathan Turner's *Re-enactor* series (2006) shows a beaming man dressed as a German soldier from the Second World War, complete with jackboots and helmet (illus. 102). Rather than photograph the man in naturalistic 'action' in the midst of a re-enactment, Turner has depicted him standing to attention in his own living room. The deadpan square-format composition frames 'Stuart' among the suburban decor of wood-effect sideboard and fringed pouffe; the butt of his rifle rests not in the dirt of the battlefield, but the floral motifs of a fitted carpet. Turner's gently humorous approach raises the troubling question of why someone would want to dress as a Nazi soldier as a leisure pursuit.

The UK artist and photographer Melanie Friend has consistently made work that questions the far-reaching consequences of war and cultures of militarization. *The Home Front* (2009–12) documents the military and civilian air displays that often take place over coastal Britain. Among other motivations, air shows provide the opportunity for commemoration of aerial dogfights such as the Battle of Britain (1940), and feed national nostalgia for 'war birds' such as Spitfires and Lancaster bombers (illus. 103). There is a

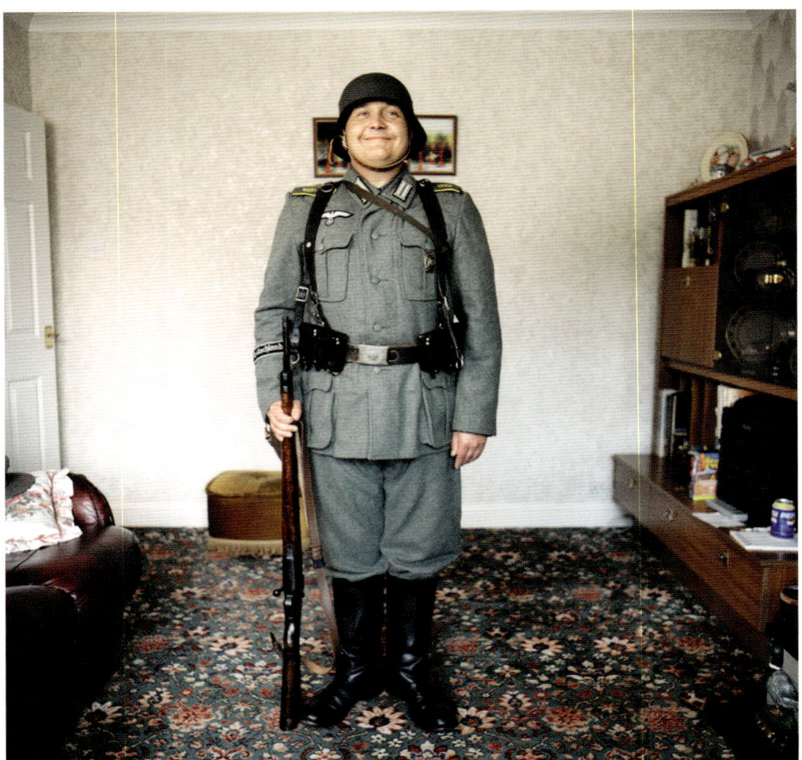

102 Jonathan Turner, *Name: Stuart. Re-enactment:* WW2 *German, Grossdeutchland 2nd Recon. (1939–1945). Location: at home in Macclesfield*, 2006, C-type print.

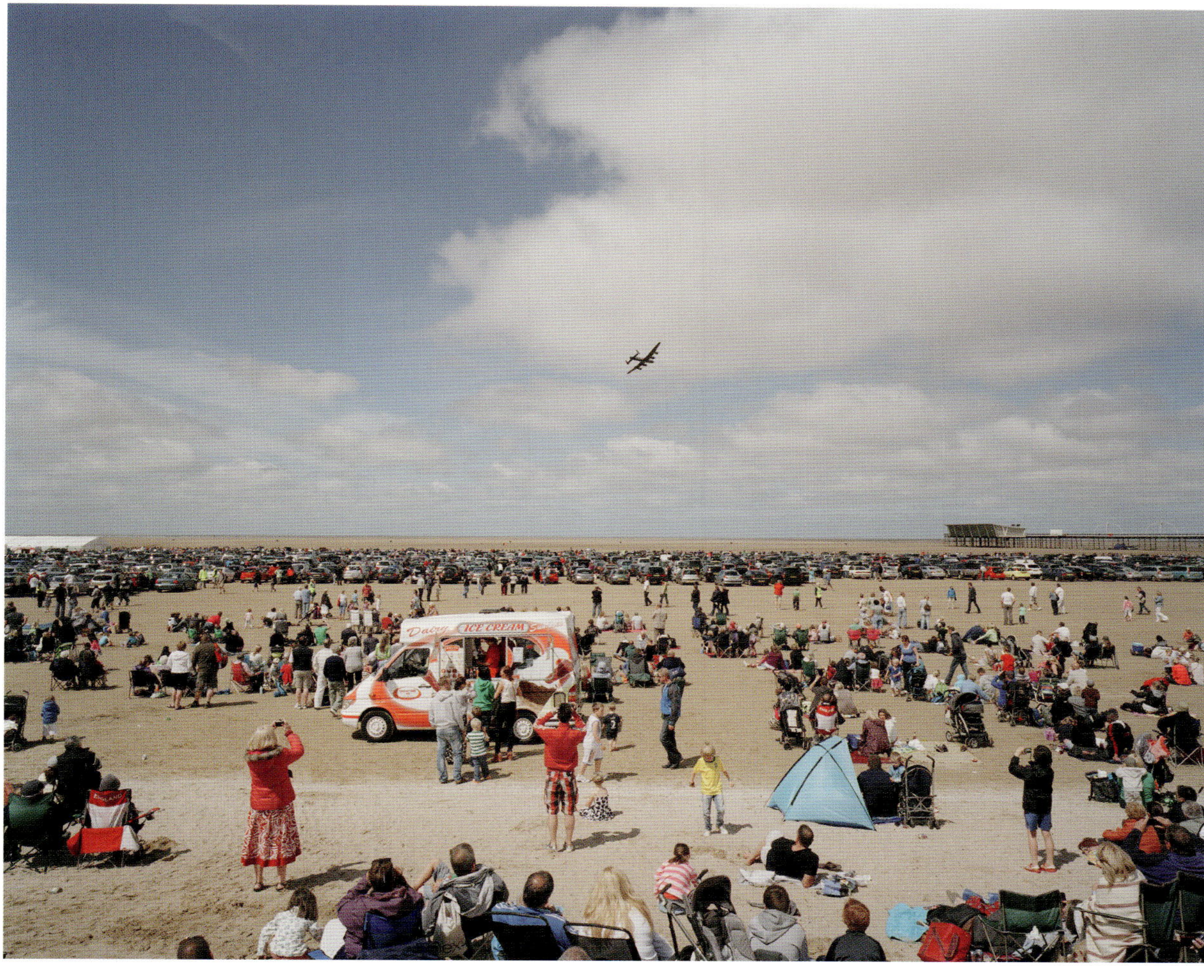

103 Melanie Friend,
Avro Lancaster Bomber (Part of the Battle of Britain Memorial Flight), Southport Beach, Merseyside, 24 July 2011, C-type print.

clear disconnect between such spectacles and the destructive capabilities of aerial warfare: promotional material for the shows, unsurprisingly, makes no mention of the hundreds of thousands of people killed by Allied air raids on Germany in the Second World War. For some onlookers, however, the association is clear. Friend's project was catalysed in part by the reaction of a traumatized four-year-old Kosovan refugee who mistook a Red Arrows display in the English seaside town of Southport for the kind of military air strike she had recently experienced in her homeland.[11] Some of Friend's images show children protecting their ears from the roar of the engines, while others seem to be running from invading fighter planes. *The Home Front* highlights the fine line between commemoration and entertainment, and reminds the viewer that patriotic displays should not be divorced from the real cost of war.

The former sites of war have long held appeal for visitors.[12] During and immediately after the First World War, photographic guidebooks were published to facilitate visits to the battlefields in which soldiers had fought or died. Books such as the *Michelin Guide to Ypres* (1919) offered the visitor comprehensive surveys of the key sites of conflict, featuring photographs of the terrain alongside explanatory text and maps. Yet the sites of former wars are continually changing, often out of all recognition, while the number of people with first-hand experience dwindles. As time passes, the viewer must possess a great deal of vision to 'see' the past – a facility that cultural historian Kitty Hauser calls 'the archaeological imagination' – or failing that, a skilled interpreter.[13]

Rose E. B. Coombs, a former radar operator and archivist at the Imperial War Museum, was a leading authority in this regard. Coombs exhaustively mapped and photographed the sites of battles including Ypres, the Somme, Mons and Verdun, as well as graves, memorials and museums. Her book *Before Endeavours Fade*, first published in 1976 and repeatedly revised, has become the indispensable guide for veterans and their descendants.[14] Coombs's photographs are descriptive rather than artistic, serving as functional visual markers for long-passed events (illus. 104). A 1970s caravan, parked at the edge of a murky pool of water, marks the spot where George v viewed the battlefront in 1917; a group of willow trees and wire fencing overlooks a flooded mine crater at Kruisstraat which today provides 'excellent

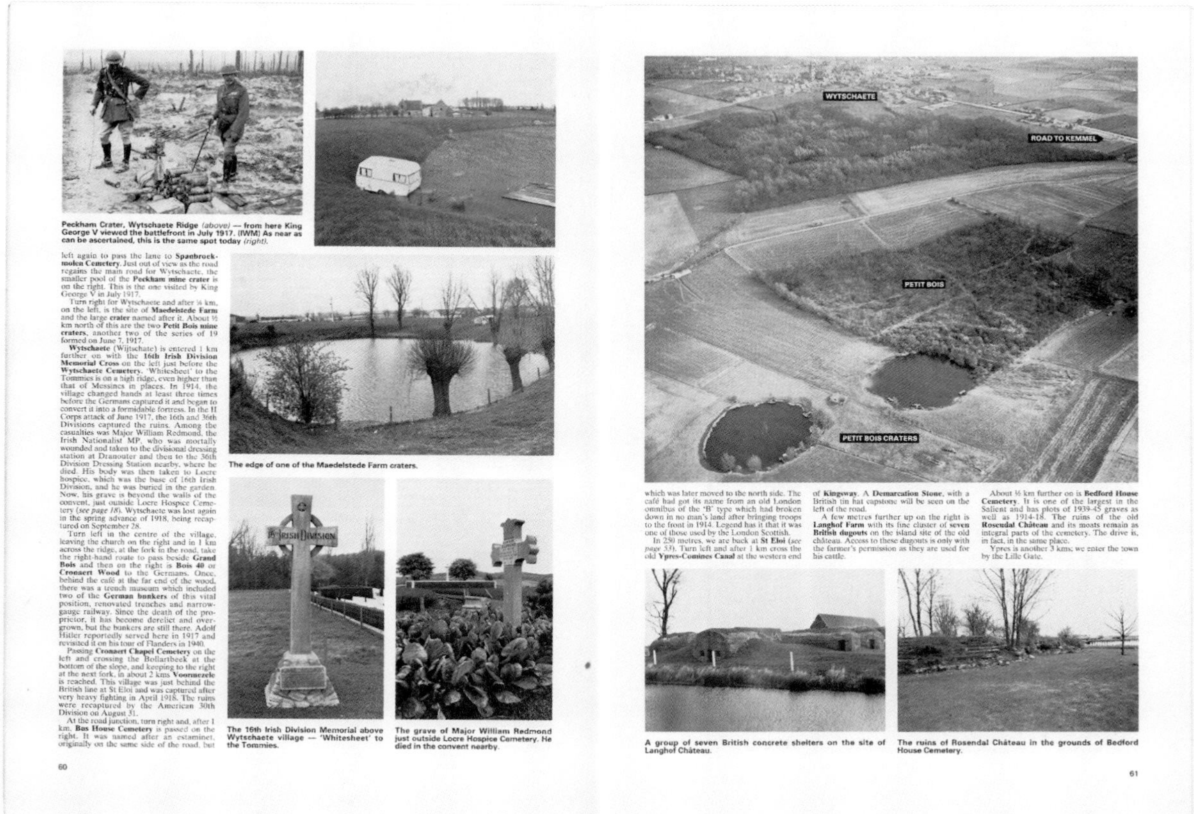

104 Rose E. B. Coombs, 'Ypres Salient Battlefield Tour: Route IV, the South-eastern Sector', from the publication *Before Endeavours Fade: A Guide to the Battlefields of the First World War* (1994).

fishing'. Coombs's book reminds the viewer that the traces of war are to be found all around us, if only we know how and where to look, and that even the most mundane location can be imbued with significance.

The ethical dimension of 'war tourism' is, like re-enactment, a complex issue. Advocates claim that it is essential to remember the past if not to repeat its mistakes; detractors argue that sites of suffering may become trivialized spectacles, entertainment for those seeking vicarious thrills of horror. The contemporary photographer Léa Eouzan has made a study of the afterlife of former concentration camps in Germany, France and Poland.[15] Her series *Histoire(s) contemporaine(s)* (Contemporary History/ies, 2006–12) records the quotidian details of sites such as Auschwitz, now equipped to serve visitors

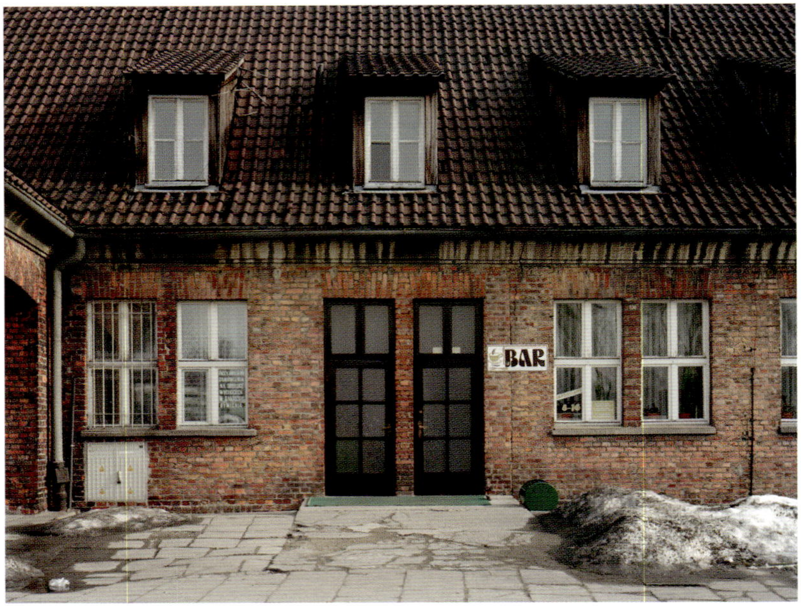

105 Léa Eouzan, *Bar – original block where there was a circuit of showers for prisoners already registered, Auschwitz Museum, Oswiecim,* March 2006, C-type print.

with car parking, toilets and a hot dog stand. One image depicts a dirty pile of snow outside a mundane brick building, a sign indicating its present-day function as a bar (illus. 105). The caption to the image, however, invites the viewer to look more closely: 'Bar – original block where there was a circuit of showers for prisoners already registered'. Knowledge of the building's prior function gives a new and sinister significance to its double doorway, which presumably separated male and female prisoners. The association, moreover, with the gas chambers that operated under the guise of showers adds to the incongruity of taking refreshments in such charged surroundings. While Eouzan implicitly critiques the 'museification' of the former camps, her images are a reminder that such places do not exist solely in a nightmarish but ungraspable past, but remain palpably present.

Ruins and Traces

Photography has a particular facility for capturing a profusion of details of material surfaces, which have long made it an apt medium for recording the physical traces of the past in the present.[16] Mid-nineteenth-century photographers including Roger Fenton (see illus. 4), Felice Beato (see illus. 12) and Harriet and Robert Tytler (see illus. 19) produced war views that portrayed ruined buildings and the detritus of fighting. In part, these images were shaped by the limitations of available technology, which precluded spontaneous images of action. However, they also serve a metaphorical function. In the Western painting tradition, the ruin exudes melancholy, inviting rumination on the transience of life and the folly of men. Nowhere is this more pointed than in the context of war, when ruination is wrought by humankind rather than the ravages of time.

The UK photographer Olive Edis was deeply moved by the effects of bombardment and dereliction in Flanders and northern France, which she witnessed while undertaking a commission by the Imperial War Museum to record women's work in the auxiliary services during the Armistice period of the First World War.[17] Her photographs of devastated towns and blasted landscapes went beyond her official remit, and seem to have been made in a private capacity. One image, depicting a bleak terrain pierced by stunted trees and shattered masonry, carries the stark one-word title *WAR*, suggesting the universal senselessness of armed conflict (illus. 106). Edis's photographs form part of a much larger photographic tradition that visually contemplates the ruins of war, from George N. Barnard's *Photographic Views of Sherman's Campaign* (1866), the product of a tour of former American Civil War battlefields, to Simon Norfolk's much more recent *Afghanistan: Chronotopia* (2002), a series of vistas bathed in golden twilight that evoke the classical ruins in landscapes by French painter Claude Lorrain.[18]

Psychological, as well as physical, aftermath is the subject of *Malvinas: Portraits and Landscapes of War* (1994–2007) by the Argentine photographer Juan Travnik. The project explores the legacy of the disastrous 1982 conflict between Argentina and the UK over the sovereignty of the Malvinas/Falklands Islands. Fighting under the command of the collapsing military junta, the poorly equipped Argentine soldiers struggled in freezing conditions, and were unprepared for the well-drilled professional British forces. Travnik made portraits of former soldiers, of whom many had been young conscripts, and travelled to the islands to photograph the detritus of war that litters the landscape. Empty crates of British munitions are tumbled in the grass, and the wreck of an Argentine Air Force Skyhawk, erroneously shot down by its own side, evokes the remains of a decomposing beast (illus. 107). The ignominy of defeat meant that the conflict has been strategically forgotten in national discourses of history and remembrance.[19] Travnik's project serves to recognize the hundreds of young lives lost, and the service of men who have been so little commemorated by their country.

106 Olive Edis, *WAR*, 1919, silver gelatin print, mounted.

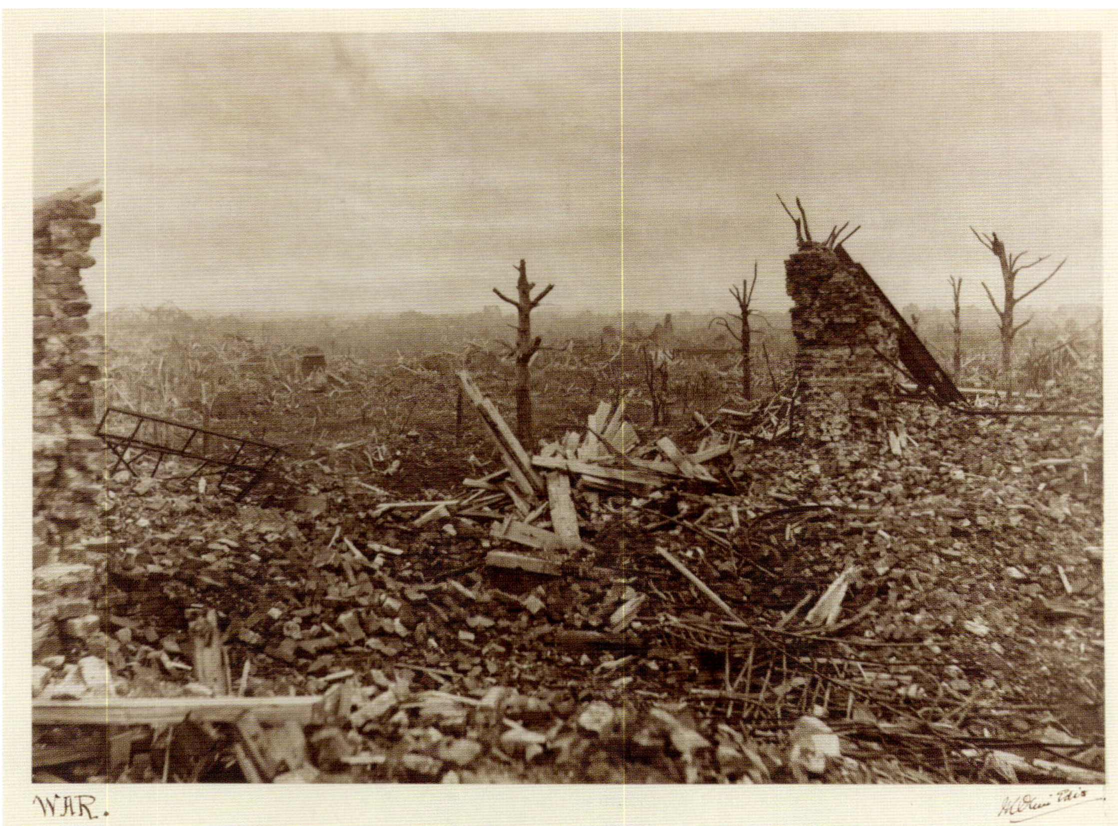

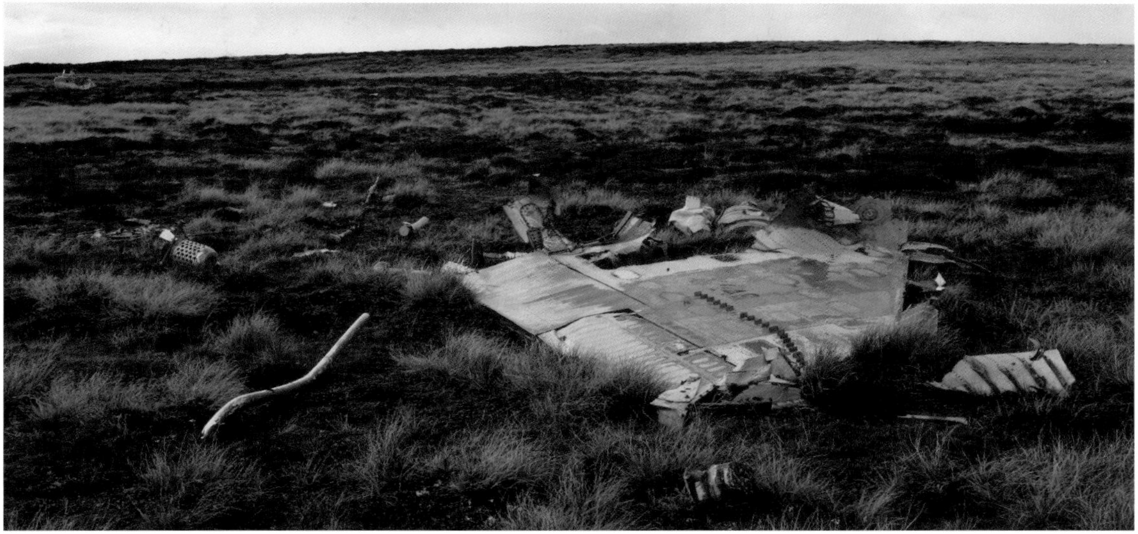

107 Juan Travnik, remains of Capt. Fausto Gavazzi's plane, Pradera del Ganso (Goose Green), Malvinas/Falkland Islands, 2007, silver gelatin print.

One of the most toxic legacies of war – both metaphorically and literally – was caused by the atomic bombs dropped on Japan in 1945. As discussed in the previous chapter, images of the immediate impact were suppressed by the U.S. authorities until the Security Treaty came into effect in 1952. Subsequently, however, photography emerged as a crucial means to communicate the disturbing after-effects, both within Japan and abroad. Presaging the 'aftermath' approach popular with present-day photographers, a wave of independent photobooks and magazines emerged in the 1950s and '60s: *Hiroshima* by Shigeichi Nagano (1952); Ken Domon's *Hiroshima* (1958); Shomei Tomatsu and Domon's *Hiroshima-Nagasaki Document 1961* (1961); Kikuji Kawada's *The Map* (1965); and Tomatsu's *11.02 Nagasaki* (1966), among others.[20]

The nature of atomic warfare meant that split-second explosions were followed by years of repercussions, visualized by the photographers via recurrent motifs of traces, scars, residues and ruins. Images depict the blood-like stains on the ceiling of the Hiroshima Peace Memorial, a Coke bottle melted by the intense heat or the pattern of a kimono burned onto the flesh of its wearer. The lingering physical effects of radiation on people – keloids, leukaemia and blindness – were recorded by Ken Domon at

Hiroshima Atomic Bomb Hospital; his photographs of twin blind girls are particularly affecting.[21] Blindness is a particularly apt metaphor, recalling at once the atomic flash, its devastating effects on victims, and the 'blindness' astutely cultivated by state agencies wishing to deny the 'atomic plague' of radiation.[22] Domon's work bears comparison with that of Philip Jones Griffiths, the Welsh photojournalist who produced a seminal account of the Vietnam War with his book *Vietnam Inc.* (1971). Jones Griffiths remained committed to documenting the far-reaching consequences of that conflict. *Agent Orange: 'Collateral Damage' in Viet Nam* (2003), named after the defoliant chemical used by the U.S. military, is an exposé of the devastating effects of herbicidal warfare and the deformities suffered by children born after the war (illus. 108).[23]

'Aftermath photography' is a relatively new movement in contemporary art and photographic practice. Numerous projects have helped shape and

108 Philip Jones Griffiths, 'Vietnam. Minh Vin Gung, born in Long An Province in 1993', from the publication *Agent Orange: 'Collateral Damage' in Viet Nam* (2003).

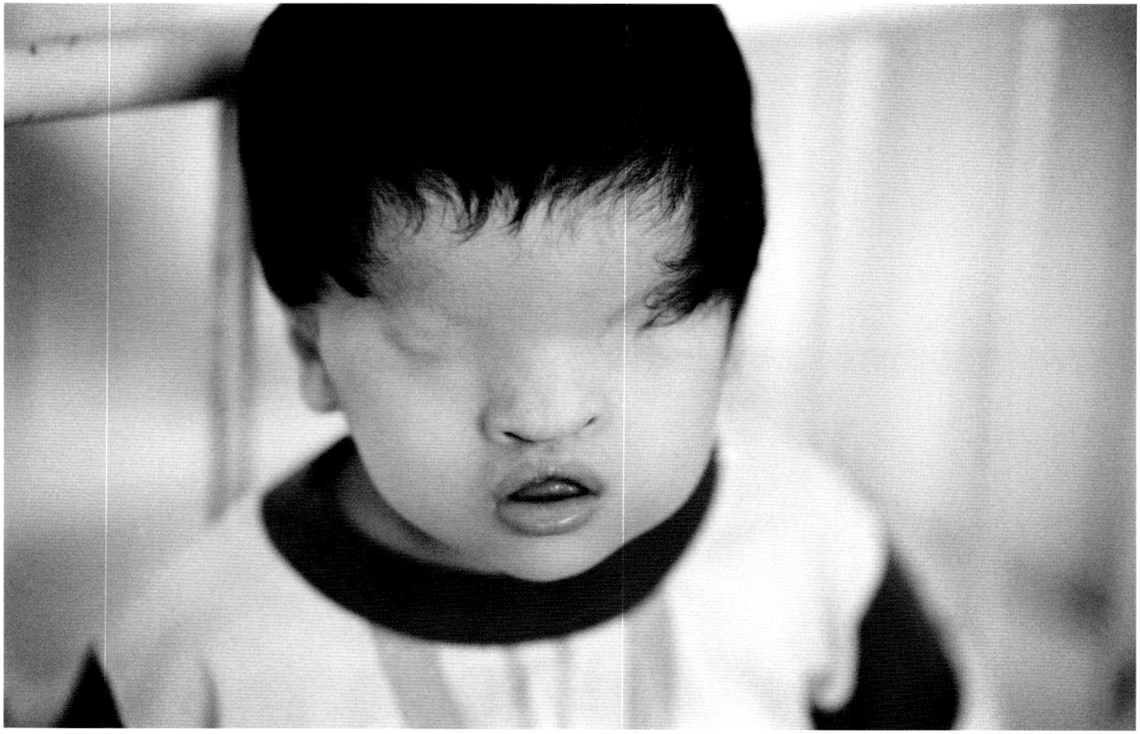

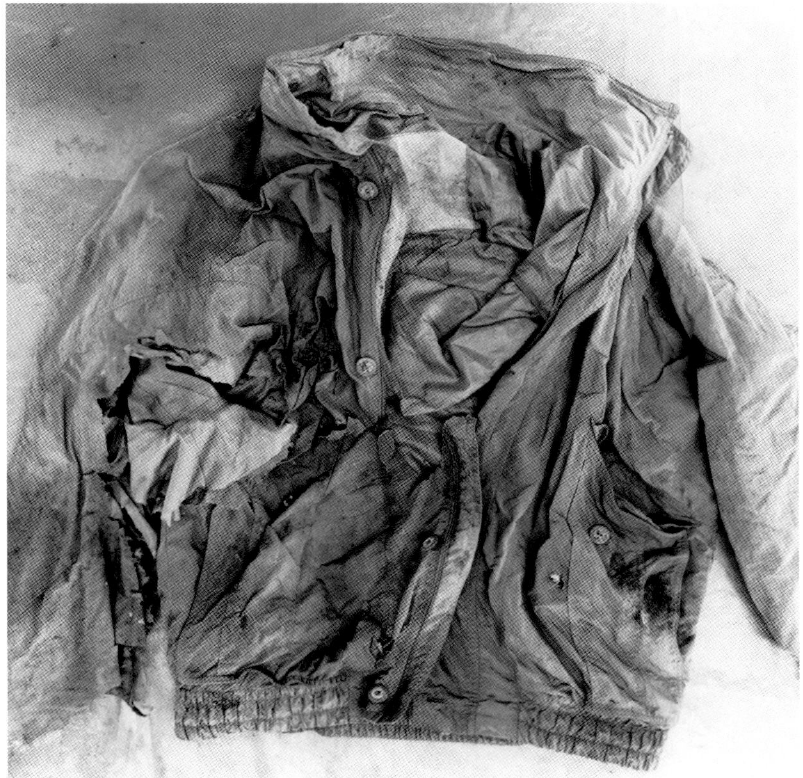

109 Frauke Eigen, from the series *Fundstücke Kosovo* (Found Objects Kosovo), 2000, silver gelatin print.

give prominence to this approach, such as the commissioning programme *The Aftermath Project: War is Only Half the Story*, established by U.S. photographer Sara Terry in 2003; the anthology *Afterwards: Contemporary Photography Confronting the Past* (2011) by Natalie Herschdorfer; and the exhibition *Conflict, Time, Photography* at Tate Modern in London, 2014.[24] Aftermath photographers make oblique visual references to the traces of past violence rather than explicit depictions, and produce extended projects rather than singular dramatic images favoured by the news media. The aim is to go beyond sensational and perhaps one-dimensional images, to instead offer more nuanced accounts and widen the panorama of those affected. Many of the contemporary projects already examined in this chapter – as well as some discussed earlier, such as those by Paula Luttringer (see illus. 79) and Seivan M. Salim (see illus. 87) – adopt the aftermath approach to fruitful effect.

The German photographer Frauke Eigen made a haunting series of images in the wake of atrocities during the Kosovo War (1998–9), when the remains of victims were exhumed from mass graves.[25] In her series *Fundstücke Kosovo* (Found Objects Kosovo, 2000–2002), she chose not to photograph the bodies themselves, but to record the victims' clothing and personal effects (illus. 109). The title suggests an archaeological approach that is endorsed by the functional black-and-white aesthetic. Yet there is an undeniable humanity and pathos to Eigen's images. The garments take on a deflated, ghostly appearance, standing proxy for the absent wearer. They seem like shedded skins, retaining – as do photographs – what Roland Barthes has characterized as the indexical imprint of the absent person.[26] Eschewing gore for metaphor, Eigen's series invites contemplation and mourning.

In other cases, the traces of war may be less visible. *Shot at Dawn* (2012–14) by UK artist Chloe Dewe Mathews explores a little-known episode in the history of the First World War: the execution of soldiers from the British, French and Belgian armies who were shot by their own side for cowardice and desertion. The histories of these men have only come to light belatedly, in part because of restrictions to files imposed by the authorities concerned, and in part owing to shame and silence on the part of some families. Around a thousand men were court-martialled and executed by firing squad between 1914 and 1918, and it is likely that at least some of these men were suffering from psychiatric illness brought on by the horrors of trench warfare.[27]

Dewe Mathews spent two years researching and photographing 23 locations at which the soldiers were shot or held in the period prior to execution. All the photographs are seasonally accurate and were taken as close as possible to the time of day at which the executions occurred, usually at daybreak. These quiet, even eerie, landscapes are very far from the dramatic combat action pictures of conventional war photography. The images are powerful precisely because no violence is explicitly shown: imagination is required to fill in the blanks of the last moments of these men ritually shot by their own comrades. A snow-covered forest of slender trees, some of them fallen to the ground, evokes the site where four North African

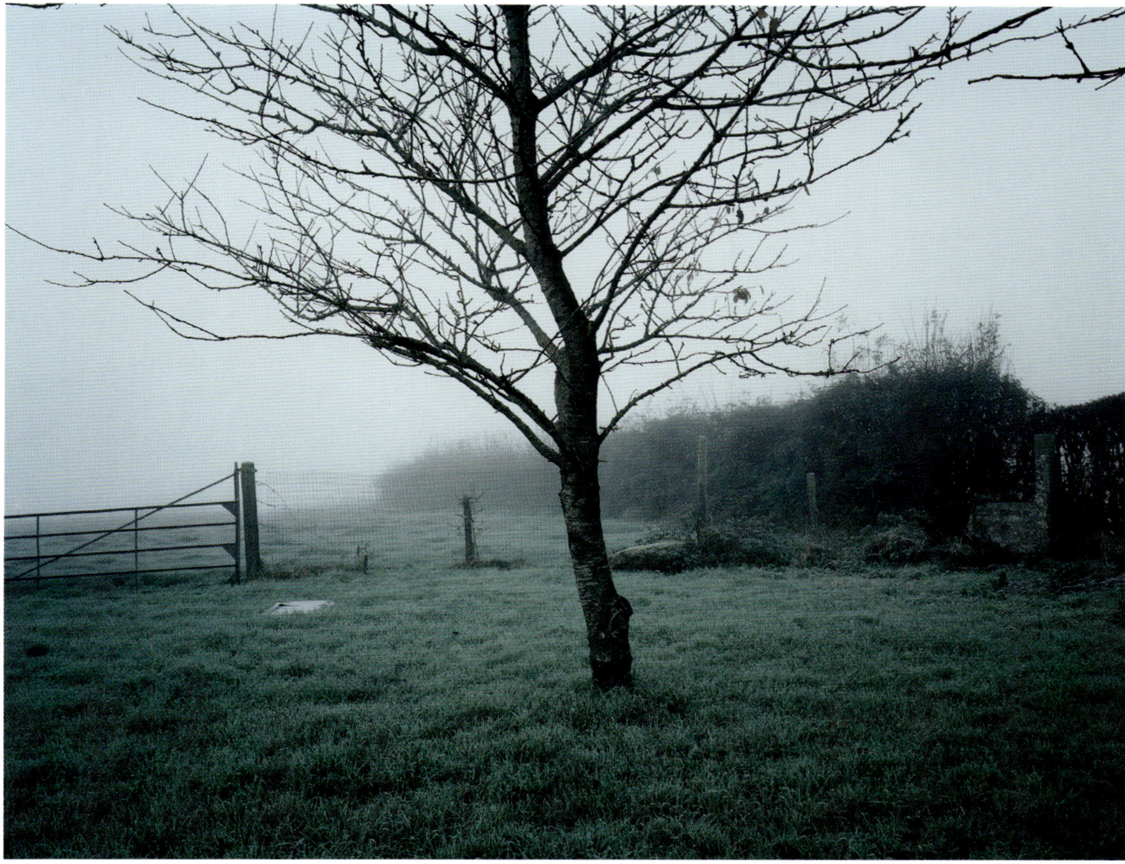

110 Chloe Dewe Mathews, *Private Joseph Byers, Private Andrew Evans, time unknown / 6.2.1915. Private George E. Collins, 07:30 / 15.2.1915. Six Farm, Loker, West-Vlaanderen*, from the series *Shot at Dawn*, 2012–14, C-type print.

soldiers were shot on 15 December 1914. A ditch filled with rainwater, marking the spot where Private Henry Hughes was killed, suggests the remains of a trench. A tree depicted in the cold early morning mist indicates the place where three British soldiers met their end in February 1915 (illus. 110). Made a hundred years later, Dewe Mathews's poignant images show places forever marked by traumatic events.

While Dewe Mathews's landscapes depend upon the viewer's imagination to reconstruct past events, the U.S. photographer Susan Meiselas offers a more explicit approach. *Reframing History* (2004) revisits her seminal photographs of the Sandinista Revolution in Nicaragua, 1978–9. Returning

25 years later, Meiselas worked with local communities to install photographic banners in the places where the images had originally been made, and filmed and photographed people's reactions.[28] For those who lived through the war, Meiselas's images evoked numerous memories: of repression and violence, euphoria and sadness.[29] Meiselas's intention was also to engage young people who did not directly experience the Revolution, although they have inherited its collected memories and its political legacies.[30] Video footage records two small boys transfixed by the near-mythic *Molotov Man* (1979), depicting Pablo 'Bareta' Arauz throwing a home-made bomb in Estelí in the final offensive, which became the icon of national liberation. In another instance, a photograph of a banner in situ shows a woman with a digital camera taking a picture of three youngsters, perhaps her children (illus. 111). The banner depicts Anastasio Somoza Portocarrero, the son of the soon-to-be-overthrown

111 Susan Meiselas, 'Anastasio Somoza Portocarrero with recruits of the elite infantry training school (EEBI)', July 1978, from the series *Reframing History*, July 2004, colour photograph.

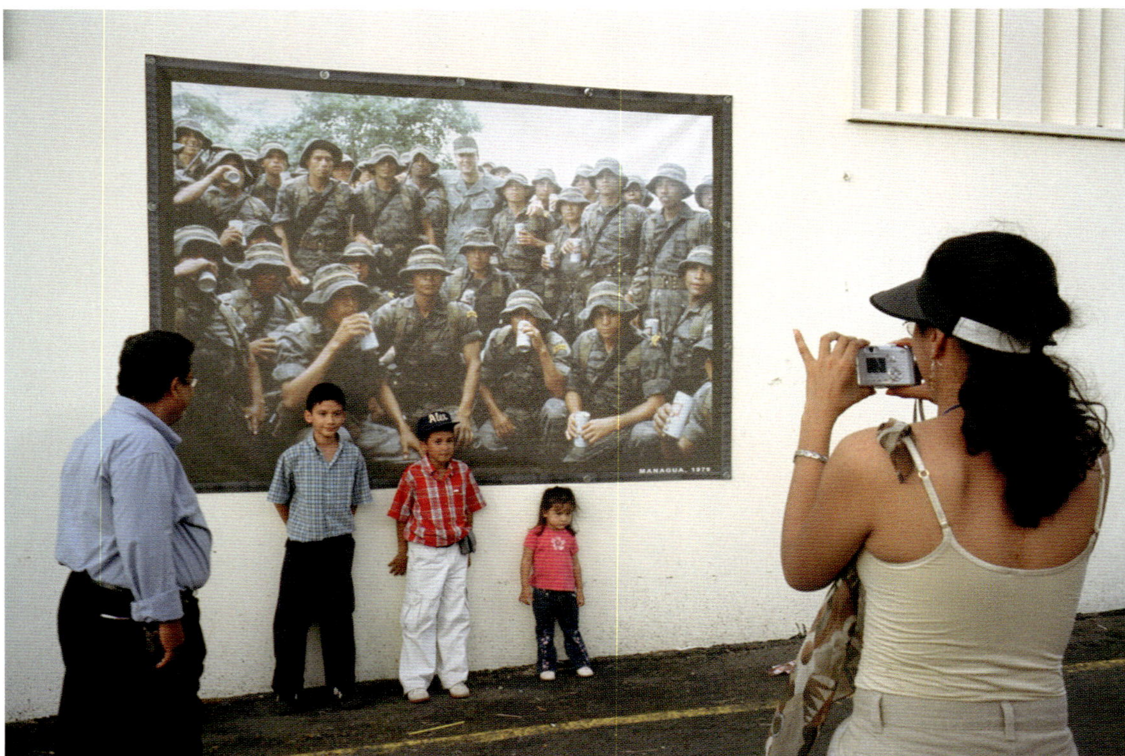

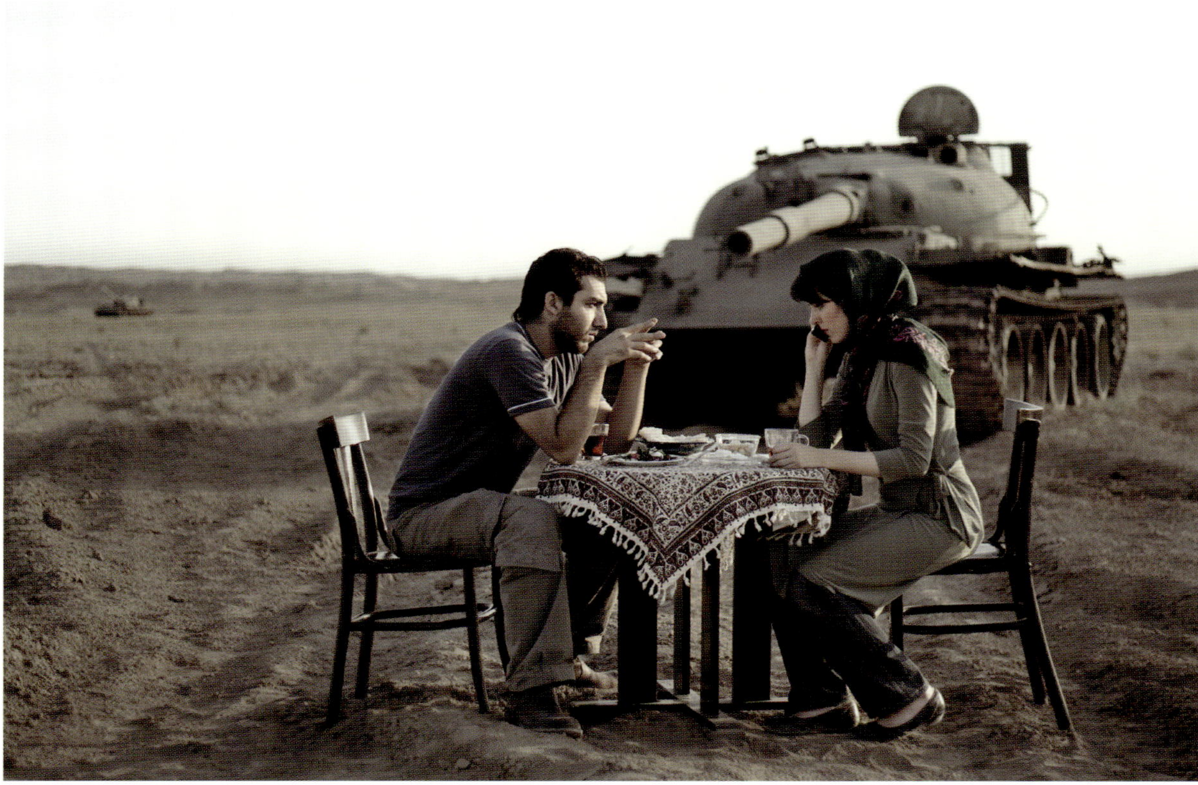

112 Gohar Dashti, 'Untitled', from the series *Today's Life and War*, 2008, C-type print.

president, and military graduates celebrating with Schlitz beer in the tradition of the U.S. Army: a metaphor for American support of the Somocista dictatorship. Meiselas's multilayered image – a photograph of a photograph being photographed – bridges past and present, inserting the new generation into a national history of the revolution and its ongoing story.

The Iranian photo-artist Gohar Dashti takes an imaginative approach to conveying the porous boundaries between past and present. *Today's Life and War* (2008) is a series of surreal images in which a couple go about their day-to-day lives: eating breakfast, hanging out washing, watching television or sleeping. Their private life, however, is transported into the terrain of a (fictionalized) battlefield. In one image, the couple eat a meal at their kitchen table, the woman embroiled in a conversation on her mobile phone, while the gun-barrel of a tank is trained on the domestic scene (illus. 112). For Dashti, who grew up during the Iran–Iraq War, the series represents the ways in which the war and its legacy continue to permeate society, and how conflict continues to infiltrate the safe space of the home through the media. In Dashti's staged images, the dramatic events of war become everyday, while ordinary life is never far from danger. The everyman and everywoman of her pictures go about their lives with stoicism, but are nonetheless haunted by the shadow of violence.

Healing, Recovery and Reconciliation

The word 'aftermath' derives from archaic English rural dialect, and denotes a crop or harvest that springs up after the first mowing. In this sense, aftermath is suggestive not simply of the repercussions of violence – the grass cut down by the scythe – but also of the struggle to revive in its wake: the growth of new shoots. War's aftermath, of necessity, is invariably a time of recovery, of reconstruction and renewal. Though the rupture of conflict may be painfully felt, it may also give rise to change and regrowth, prompting creativity and productivity. Photography has an important, albeit overlooked, role to play in this stage. This chapter, and the book, closes with a discussion of some of the ways in which photography might support healing, reconciliation and even forgiveness.

Some photographers have chosen to record processes of physical recovery, emphasizing the potential for healing and transformation in the aftermath of war. Phan Thi Kim Phuc, or 'the girl in the picture' as her biography is titled, is better known as the Vietnamese child suffering napalm burns in Nick Ut's iconic 1972 image from the Vietnam War (see illus. 14).[31] After having worldwide attention thrust upon her, Phuc established a foundation to assist child victims of war and in 1994 became a UNESCO Goodwill Ambassador, advocating forgiveness in the face of hatred. In 1995 Joe McNally photographed her for *Life* magazine, on the occasion of the birth of her son, Thomas. McNally's portrait is a classic of maternal love, recalling the Western art tradition of the Virgin and Child (illus. 113). What distinguishes the image, however, is the scarred tissue of Phuc's body: her back, shoulder and arm criss-crossed with the marks of old lesions. The possibility of recovery and renewal, symbolized by both the healed skin and the newborn infant, points to the possibility for reconciliation after the violence of war. Yet there is perhaps also an effort (unconscious or otherwise) to overwrite the searing original image that stood for the brutality of the U.S. campaign in Vietnam, and which naturalizes its recovery.

In recent years, photography has been used in a therapeutic capacity to help people come to terms with their experiences of war. Photo-therapy

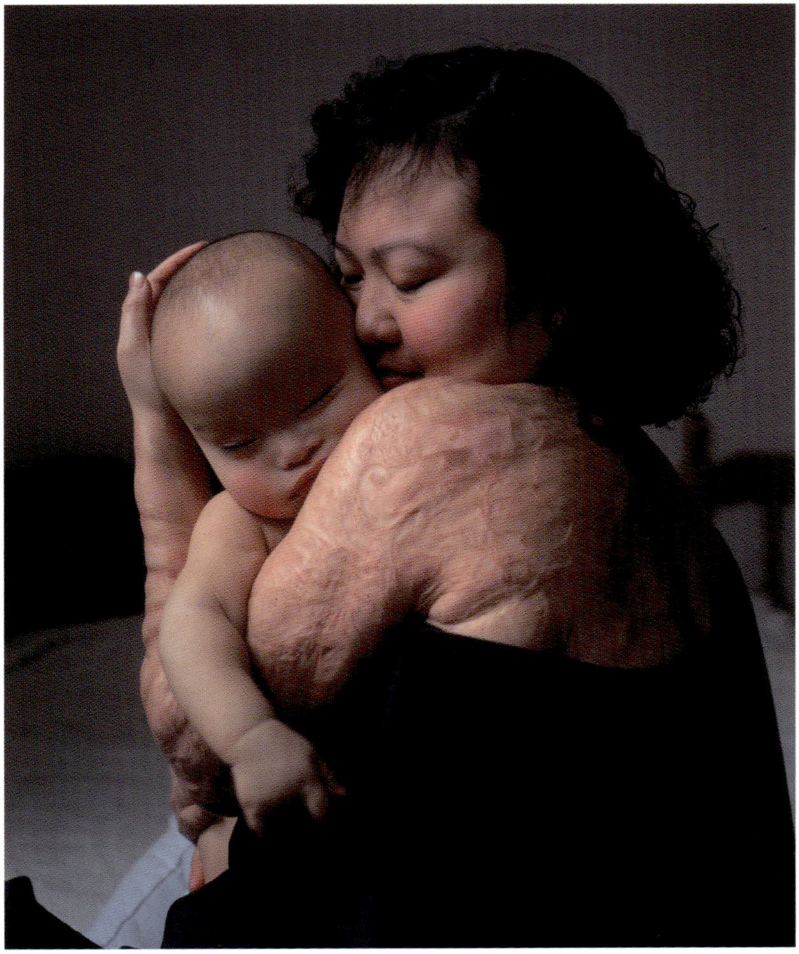

113 Joe McNally, 'Kim Phuc and son', 1995, C-type print.

emerged as an artistic and medical practice in the 1980s, in part through the seminal work of feminist photographers Jo Spence and Rosy Martin.[32] The practice provides a theorized framework for making or manipulating photographic images in order to resolve trauma and to communicate beyond words. As well as being undertaken at military rehabilitation centres, independent projects have enabled veterans to take part outside official structures. The U.S. artist Monica Haller was inspired to set up the Veterans Book Project in 2010 after collaborating with Riley Sharbonno, an Army Nurse at the notorious prison in Abu Ghraib, Iraq, on the book *Riley and his Story* (2009).[33] Haller subsequently devised workshops for participants, both military and civilian, encouraging them to bring

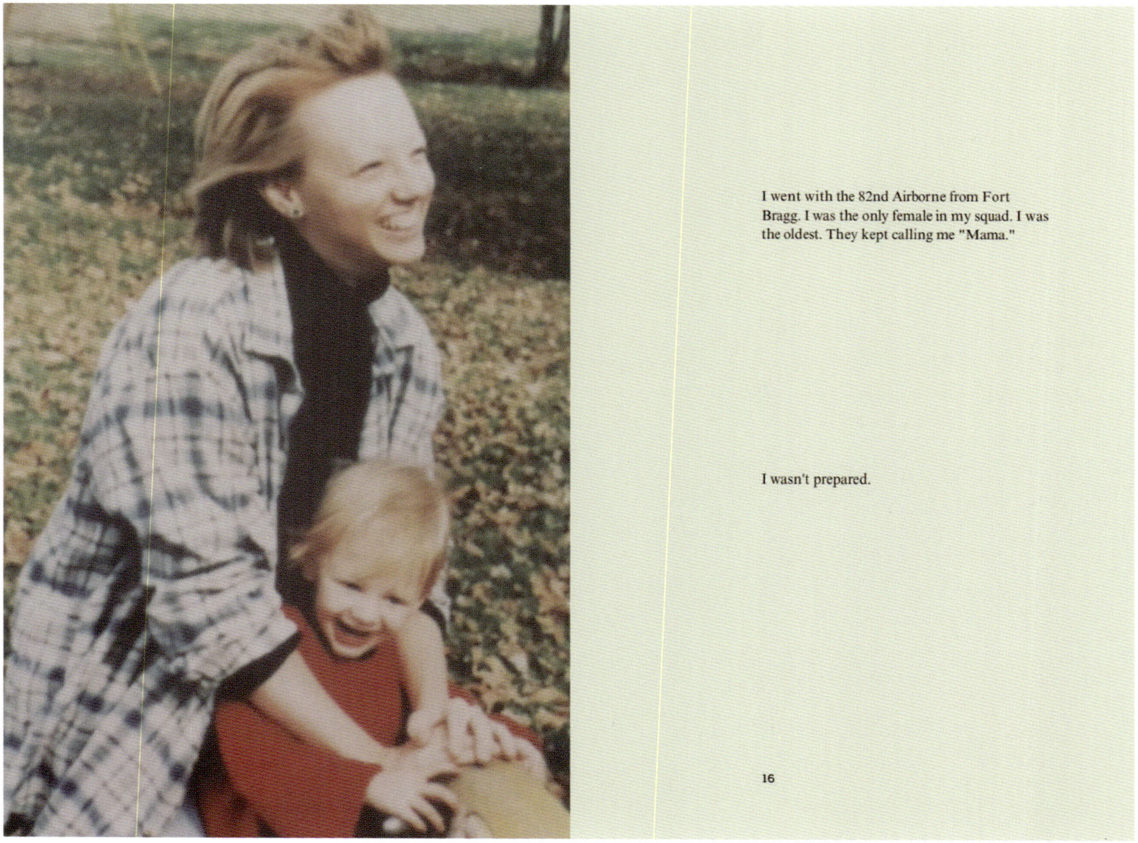

114 Juliet Madsen, double-page spread from a photobook made at Veterans Book Workshop, Colorado Springs, 2011.

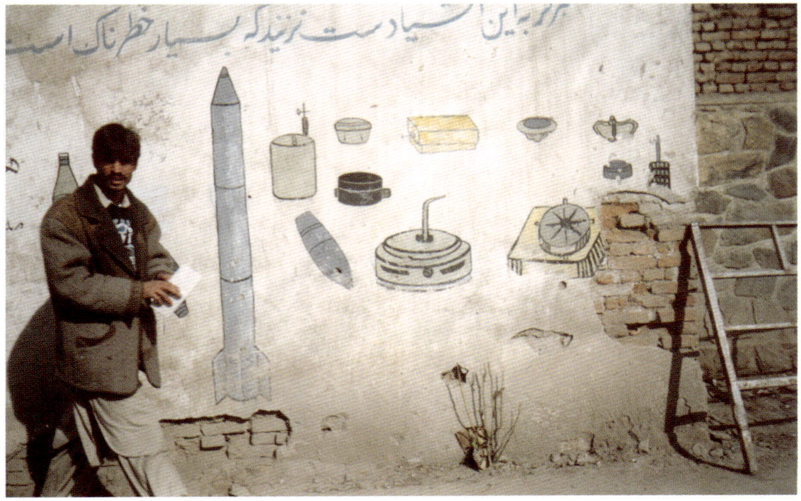

115 Nabiela, 'Warnings are painted on the walls to tell people to take care of mines . . .', from the PhotoVoice project *Shooting Kabul*', Afghanistan, 2002, colour photograph.

photographs to be edited and sequenced to form meaningful narratives.[34] As well as helping participants to better understand their experiences, the books can be helpful in the first steps to communicating with others. Juliet Madsen, a paramedic at an internment camp in Iraq, made her book for her husband and children as well as herself, declaring, 'I want them to see and understand that I really do realise that I am not who I was' (illus. 114).[35] Her book offers an honest and moving account of illness, bomb trauma and the guilt she feels as a mother who has worked in a war zone.

Community photography projects have delivered benefits to post-conflict societies. The 'photovoice' concept, which evolved from the health education research of Caroline Wang and Mary Ann Burris in the 1990s, offers a photography-based methodology to engender discussion and dialogue among communities in a way that is useful to them.[36] Through structured and facilitated workshops, participants have the opportunity to express their viewpoints through making photographs, and to share them with others. Such projects may play a modest but nonetheless valuable role in the processes of reconciliation and peacebuilding. 'Shooting Kabul', a project run by the charity PhotoVoice in 2002, enabled young Afghan girls to make photographs reflective of their lives as the nation emerged from Taliban rule. Through their images, they discussed their views on peace and reconstruction, and

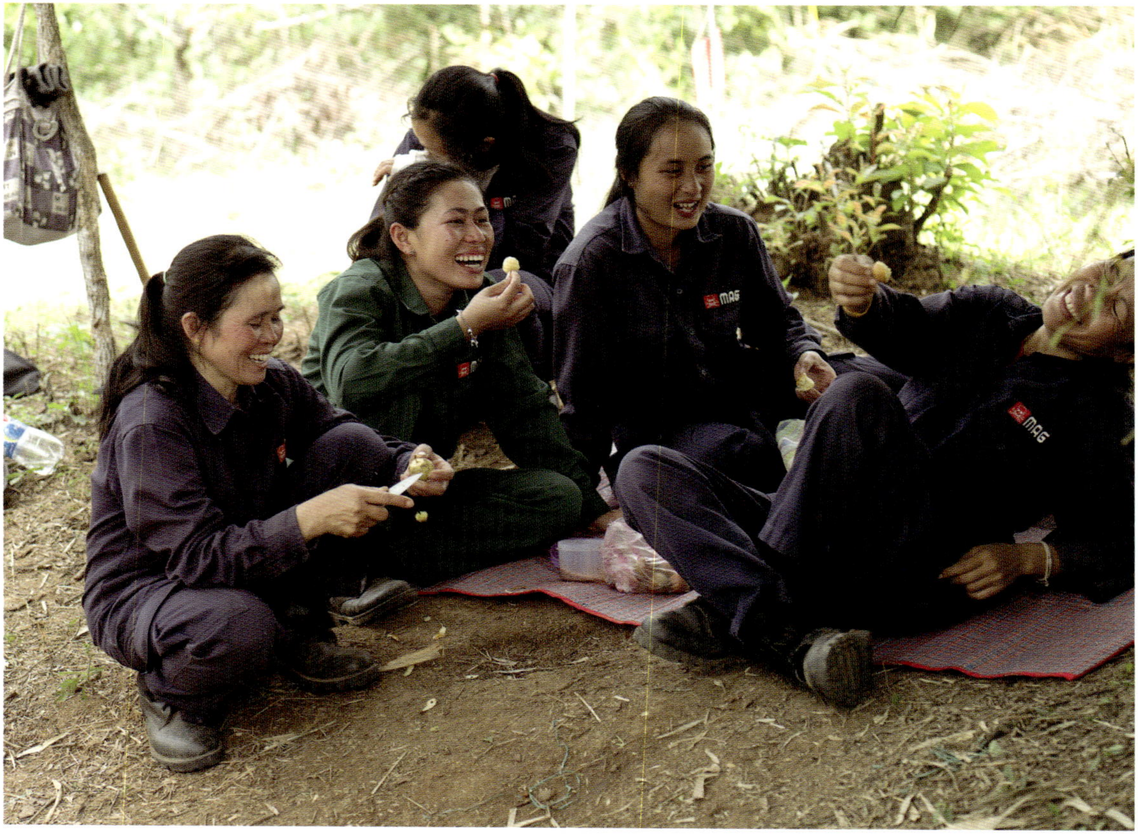

116 Tessa Bunney, 'Deputy team leader, Manixia Thor, relaxes with her colleagues from UCT6 during breaktime while clearing UXO [unexploded ordnance] in Ban Namoune', from the series *The Women of UCT6*, 2014, C-type print.

women's and working children's rights and education.[37] A photograph by a participant named Nabiela depicts one of the many vestiges of conflict: a crumbling mural warning of the dangers of anti-personnel mines (illus. 115). She explains:

> The warnings are painted on the walls to tell people to beware of mines. This is good as it means people can live more safely. My neighbour used to collect steel and he thought a mine was steel he could recycle and took it. He now cannot walk – he has lost a foot and a hand. His sister comes here to Aschiana. The landmines are found in many places outside. Small children are taught what the different mines look like so they know not to

touch them. Sometimes they are hidden. Before they used to be near our houses but now they have been cleared. Many children were injured as they used to play in mined areas.

For Nabiela, the image recalls the traumatic history of a neighbour who accidentally detonated a device, but her image is also forward-looking, conveying how future generations may live more safely through education. By supporting participants to recognize signs of transition and reconciliation, projects like 'Shooting Kabul' may help them to progress from states of trauma and anxiety to relative security. While images – like Nabiela's – may sometimes be aesthetically arresting, it is the process, not the end product, that is valuable.

While living in the People's Republic of Lao, contemporary UK photographer Tessa Bunney documented community efforts to recover from the Vietnam War. Like the citizens of Kabul, Laotians face continual danger due to unexploded ordnance that remains embedded in the land. Working on behalf of the charity Mines Action Group, Bunney recorded the activities of UCT6, a Vietnamese woman-only mine clearance team. Their arduous task involves combing the ground with metal detectors and removing live devices, but also affords opportunities for camaraderie (illus. 116).[38] Bunney records other transformations as well: some of the villagers have been enterprising in their reuse of bombs, using them as building material for fences and houses or animal troughs. One man has even built a business smelting the metal gathered from decommissioned ordnance to fabricate spoons, an extraordinary act of pragmatism and creative resilience. Bunney's photographs offer another take on the history of the Vietnam War, one which foregrounds the agency of the survivors in recovery and reconstruction.

The series *Anatomy of Forgiveness* (2014) by Dutch-Croatian photographer Lana Mesic was made twenty years after the 1994 genocide of Tutsis in the East African nation of Rwanda. Mesic worked with Association Modeste et Innocent, a community organization in Butare supporting reconciliation programmes that bring together victims and perpetrators. Mesic met seventeen pairs of former workshop participants that wished to be photographed and tell their story. In response to the challenge of photographing what she terms 'the invisibility of reconciliation', Mesic asked the pairs to remember their moment of forgiveness, and to re-enact

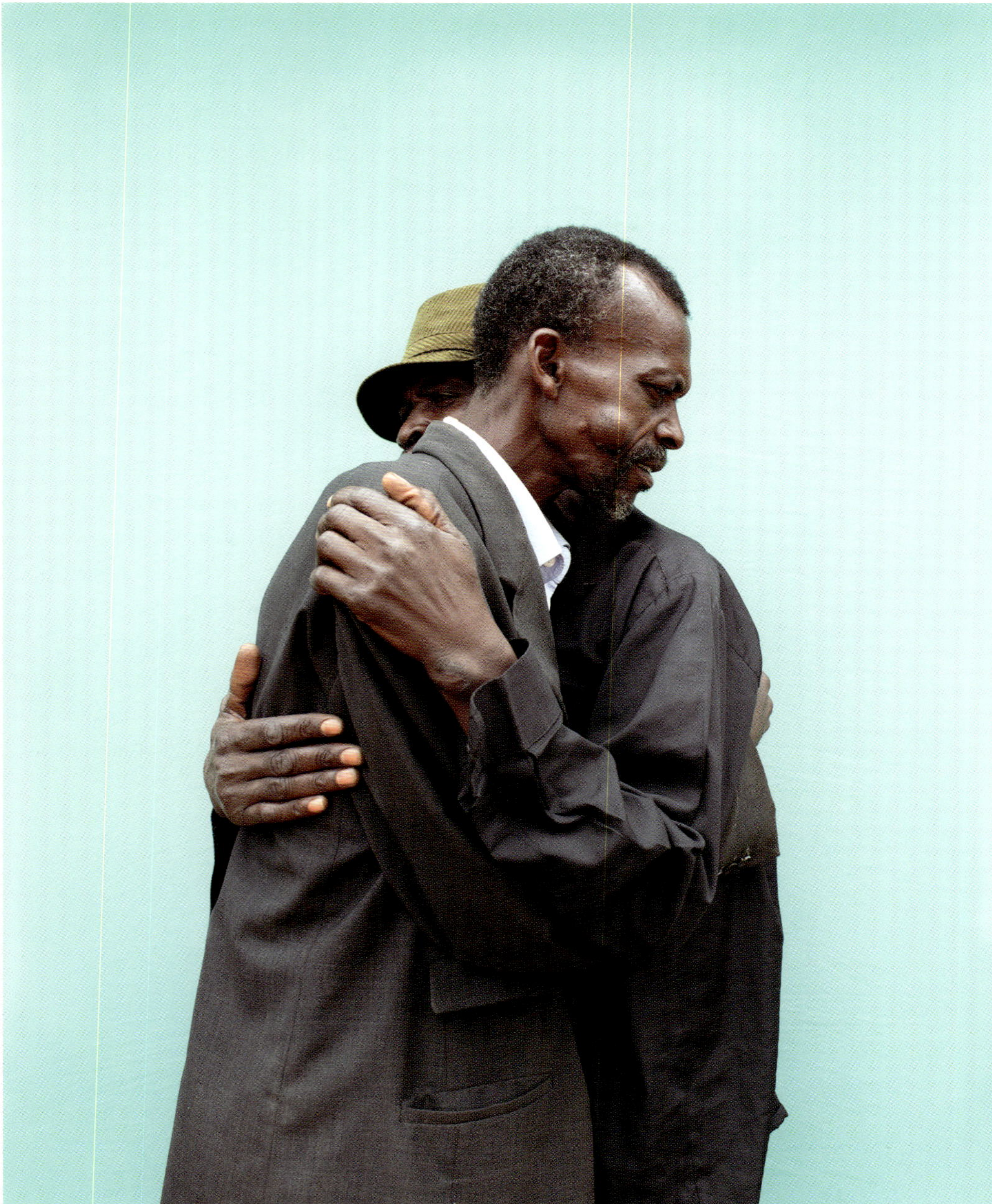

what happened. She also devised 'scales of forgiveness', marking ordinary objects such as plastic bottles and leaves with scales from 1 to 10, and asking each person to indicate how much they had forgiven each other. The resulting photographs do not attempt to naturalize the process of forgiveness, but instead foreground the artificiality of the situation and the efforts that must be made. The images of the 'scales of forgiveness' reveal disparities between victim and perpetrator with, unsurprisingly, the former much less likely to indicate that they had completely forgiven the latter. The couples themselves are pictured in makeshift studios, the staged nature of their encounters emphasized by the occasional glimpse of the edges of the sagging coloured fabric backdrops. Some couples are captured in awkward and distanced poses; others embrace as if they were reliving a genuine moment of compassion and reconciliation (illus. 117).

Anatomy of Forgiveness is both painful and moving, highlighting the complexities of peace processes, as well as the unfathomable resilience of people determined to recover after violent conflict. Such projects, however, are all too rarely seen. The Western news media economy is heavily weighted to images of tension and conflict, and images of recovery and reconciliation tend to be sidelined. While war photography is a much-lauded genre, the notion of 'peace photography' is, as yet, embryonic.[39] Positive depictions of attempts to build more tolerant and resilient communities are crucial, if we are to visualize a future in which armed conflict is lessened. Photography may yet have an important role to play in this future.

117 Lana Mesić, *Christophe and Francois*, from *Anatomy of Forgiveness*, 2014, C-type print.

References

All translations are the author's except where otherwise indicated.

Introduction

1 Exceptions to the standard approaches include Anne Wilkes Tucker and Will Michels, eds, *War/Photography: Images of Armed Conflict and its Aftermath* (New Haven, CT, 2012), thematized by stages of war; and Simon Baker and Shoair Mavlian, eds, *Conflict, Time, Photography* (London, 2014), sequenced according to time elapsed after conflict.

2 See, for instance, Antony Penrose, ed., *Lee Miller's War: Photographer and Correspondent With the Allies, 1944–1945* (London, 1992).

3 I am grateful to Peter Hamilton for sharing his interpretation of the circumstances of the photograph's production.

4 Oliver Wendell Holmes, 'Photographic Phases', *New York Times*, 21 July 1862, p. 5.

5 Oliver Wendell Holmes, 'Brady's Photographs: Pictures of the Dead at Antietam', *New York Times*, 20 October 1862, p. 5.

6 Ibid., p. 5.

7 John O'Brian, 'The Nuclear Family of Man', *Asia-Pacific Journal: Japan Focus*, VI/7 (2008).

8 Ned O'Gorman and Kevin Hamilton, 'EG&G and the Deep Media of Timing, Firing, and Exposing', *Journal of War and Culture Studies: The Business of War Photography*, IX/2 (2016), pp. 182–201.

9 Christopher Pinney, ed., *Photography's Other Histories* (Durham, NC, 2003); Ali Behdad, *Camera Orientalis: Reflections on Photography of the Middle East* (Chicago, IL, 2016).

10 The military historian John Keegan asserts warfare to be 'an entirely masculine activity', from which women, 'with the most insignificant exceptions, have always and everywhere stood apart'. John Keegan, *A History of Warfare* (London, 2004), p. 76. Cynthia Enloe contends that 'the military has to constantly redefine "the front" and "combat" as wherever "women" are not.' Cynthia Enloe, *Does Khaki Become You? The Militarization of Women's Lives* (Boston, MA, 1983), p. 15.

11 For Capa's D-Day photographs, see Richard Whelan, *This is War! Robert Capa at Work: 1936–1945* (Göttingen, 2007) and A. D. Coleman, 'Conflict of Interest, Cubed: Robert Capa's D-Day Photos, John Morris, and the NPPA' (2017), www.imediaethics.org, accessed 30 June 2018; for *Flag Raising at Iwo Jima*, see Anne Wilkes Tucker and Will Michels, eds, *War/Photography: Images of Armed Conflict and its Aftermath* (New Haven, CT, 2012), pp. 258–79; for *Reaching Out* see Liam Kennedy, *Afterimages: Photography and U.S. Foreign Policy* (Chicago, IL, and London, 2016), pp. 13–51.

one: Despatches from the Combat Zone

1 'A Lady's Postscript to a Crimean Letter', *Punch* (1855), cited in Pat Hodgson *Early War Photographs* (Reading, 1974), p. 11.

2 The earliest known images connected with war are daguerreotypes made by itinerant commercial portraitists, depicting U.S. troops in Saltillo, Mexico, during the Mexican-American War of 1846–8. However, they seem to have been private undertakings and made no public impact at the time. Olivier Debroise, *Mexican Suite: A History of Photography in Mexico*, trans. and rev. by Olivier Debroise and Stella de Sá Rego (Austin, TX, 2001), pp. 164–8.

3 John Hannavy, 'History: 4. 1850s', in *Encyclopedia of Nineteenth-century Photography*, ed. John Hannavy (New York and London, 2008), p. 685.

4 In particular, *The Times* war correspondent William Howard Russell had reported on the mismanagement of the invasion and hardships faced by soldiers. William Howard Russell, *Russell's Despatches from the Crimea, 1854–1856*, ed. Nicholas Bentley (London, 1966).

5 Helmut Gernsheim, *A Concise History of Photography* (New York, 1986), p. 64.

6 Peter Harrington, 'The First True War Artist', *MHQ: The Quarterly Journal of Military History*, IX/1 (Autumn 1996), pp. 100–109.

7 Ulrich Keller, *The Ultimate Spectacle: A Visual History of the Crimean War* (New York and London, 2001), p. 133.

8 For the perceived value of photographic 'accuracy' as compared to painting or printmaking, see Bernd Hüppauf, 'The Emergence of Modern War Imagery in Early Photography', *History and Memory*, V/1 (1993), pp. 130–51.

9 Bodo von Dewitz and Robert Lebeck, eds, *Kiosk: A History of Photojournalism, 1839–1973* (Cologne, 2001), pp. 22, 28–9.

10 'Shadows of War: Roger Fentons's Photographs of the Crimea 1855, The Queen's Gallery, Palace of Holyroodhouse: After the War', www.royalcollection.org.uk, accessed 22 July 2018.

11 Alan Trachtenberg, 'Albums of War: On Reading Civil War Photographs', *Representations*, 9 (1985), pp. 1–32.

12 'From the Seat of War: Truly Superb Photographic Views – A Picture Worth Framing', *New Orleans Daily Crescent*, 28 May 1861, p. 2.

13 Cited in Ron Field, 'Camp Photographers: Pictures by the Thousand', *Military Images*, XXXV/4 (2017), pp. 60–64.

14 Reported in *New York Tribune*, 20 August 1862, cited in Mary Warner Marien, *Photography: A Cultural History: Third Edition* (London 2010), pp. 105–8.

15 See, for example, 'Scenes on the Battlefield of Antietam – From Photographs by Mr M. B. Brady', *Harper's Weekly*, VI/303 (18 October 1862), pp. 664–5.

16 Emily Godbey, '"Terrible Fascination": Civil War Stereographs of the Dead', *History of Photography*, XXXVI/3 (2012), p. 268.

17 Trachtenberg, 'Albums of War', pp. 6–7.

18 Godbey, 'Terrible Fascination', pp. 265–74; Bob Zeller, *The Civil War in Depth: History in 3-D* (San Francisco, CA, 1997).

19 Alexander Gardner, *Catalogue of Photographic Incidents of the War, From the Gallery of Alexander Gardner, Photographer of the Army of the Potomac, Corner of 7th and D Streets Washington, DC* (Washington, DC, 1863).

20 Jody M. Seels and Barbara A. Seels, 'Civil War Photography and its Impact from 1863–1993', in *Visual Literacy in the Digital Age: Selected Readings from the Annual Conference of the International Visual Literacy Association* (Rochester, NY, 1993), pp. 382–90.

21 David Rudd Cycleback, 'Halftone Printing', in *Encyclopedia of Nineteenth-century Photography*, ed. John Hannavy (New York and London, 2008), pp. 632–3; Dewitz and Lebeck, *Kiosk*, p. 40.

22 Amanda Hopkinson and Robin Lenman, 'Photojournalism', in *The Oxford Companion to the Photograph*, ed. Robin Lenman (Oxford, 2005), pp. 488–9.

23 The original source for the legend, which was directed at a staff illustrator rather than a photographer, is probably James Creelman, *On the Great Highway: The Wanderings and Adventures of a Special Correspondent* (Boston, MA, 1901), pp. 177–8.

24 Marien, *Photography: A Cultural History*, pp. 224–6.

25 Paula E. Calvin and Deborah A. Deacon, *American Women Artists in Wartime, 1776–2010* (Jefferson, NC, and London, 2011), pp. 72–3.

26 Christopher Stolarski, 'The Rise of Photojournalism in Russia and the Soviet Union, 1900–1931', PhD Thesis, John Hopkins University (Baltimore, MA, 2013), pp. 129–32.

27 Dewitz and Lebeck, *Kiosk*, pp. 62–81.

28 Cited in John Mraz, *Photographing the Mexican Revolution: Commitments, Testimonies, Icons* (Austin, TX, 2012), p. 57.

29 The New York *Herald* correspondent was A. A. Willis, cited in George Leighton, 'The Photographic History of the Mexican Revolution', in *The Wind That Swept Mexico: The History of the Mexican Revolution, 1910–1942*, ed. Anita Brenner (Austin, TX, 1984), p. 295.

30 Andrea Noble, *Photography and Memory in Mexico: Icons of Revolution* (Manchester, 2010), pp. 25–8.

31 Paul J. Vanderwood and Frank N. Samponaro, *Border Fury: A Picture Postcard Record of Mexico's Revolution and U.S. War Preparedness, 1910–1917* (Albuquerque, 1988), p. 13; John Mraz, *Looking for Mexico: Modern Visual Culture and National Identity* (Durham, NC, 2009), pp. 60–63.

32 John Mraz, *Looking for Mexico*, pp. 63–4; Mike Dash, 'Uncovering the Truth Behind the Myth of Pancho Villa, Movie Star', www.smithsonian.com, 6 November 2012.

33 Nancy Brandt, 'Pancho Villa: The Making of a Modern Legend', *The Americas*, XXI/2 (1964), pp. 137–8.

34 Dewitz and Lebeck, *Kiosk*, pp. 108–59; Esther Gabara, *Errant Modernism: The Ethos of Photography in Mexico and Brazil* (Durham, NC, and London, 2008), pp. 120–95.

35 The advisor was Daniel Longwell, Luce's special assistant at *Time*. Cited in Loudon Wainwright, *The Great American Magazine: An Inside History of Life* (New York, 1986), p. 121.

36 'The Greatest War Photographer in the World', *Picture Post*, 3 December 1938.

37 Cited in David E. Scherman, 'Introduction: The Photographer Under Fire', in *Life at War* (New York, 1981), p. 6.

38 Caroline Brothers, *War and Photography: A Cultural History* (Abingdon, 1997), pp. 178–85; Michael Griffin, 'The Great War Photographs: Constructing Myths of History and Photojournalism', in *Picturing the Past: Media, History, and Photography*, ed. Bonnie Brennen and Hanno Hardt (Chicago and Urbana, IL, 1999); Richard Whelan, 'Robert Capa's Falling Soldier', *Aperture*, 166 (2002), pp. 48–55; J. M. Susperregui, 'Localización de la fotografía "Muerte de un miliciano" de Robert Capa', *Communication and Society*, XXIX/2 (2016), pp. 17–44.

39 'Photographic Pictures in Relief', *The Practical Mechanic's Journal*, 1854, cited in Hüppauf, 'The Emergence of Modern War Imagery', p. 130.

40 Cited in Jane Carmichael, *First World War Photographers* (London and New York, 1989), p. 60.

41 Frank Hurley, *Exhibition of War Photographs by Capt. F. Hurley, late Official Photographer with the AIF, held at The Kodak Salon, Sydney* (undated).

42 Jennifer Wellington, *Exhibiting War: The Great War, Museums, and Memory in Britain, Canada, and Australia*, (Cambridge, 2017), pp. 76–80.

43 Michael Griffin, 'The Great War Photographs: Constructing Myths of History and Photojournalism', in *Picturing the Past: Media, History and Photography*, ed. Bonnie Brennen and Hanno Hardt (Urbana, IL, 1999), pp. 122–57.

44 Dr David F. Rennie, *The British Arms to Northern China and Japan: Pekin 1860; Kagoshima 1862* (London, 1864), cited in David Harris, *Of Battle and Beauty: Felice Beato's Photographs of China* (Santa Barbara, CA, 1999), p. 29.

45 Luke Gartlan, 'Beato, Felice (*c.* 1834–1906) in *Encyclopedia of Nineteenth-century Photography*, ed. John Hannavy (New York and London, 2008), pp. 128–31.

46 Oliver Wendell Homes, 'Brady's Photographs: Pictures of the Dead at Antietam', *New York Times*, 20 October 1862, p. 5.

47 Godbey, 'Terrible Fascination', pp. 265–74.

48 Although the date is usually given as 1865; it was in fact January 1866. Anne E. Peterson, 'Alexander Gardner in Review', *History of Photography*, XXXIV/4 (2010), pp. 356–67.
49 *Accidental Napalm*, see Robert Hariman and John Louis Lucaites, *No Caption Needed: Iconic Photographs, Public Culture, and Liberal Democracy* (Chicago, IL, 2007), pp. 171–207; Michael Griffin, 'Media Images of War', *Media, War and Conflict*, III/7 (2010), pp. 7–41.
50 Tim Page and William Shawcross, *Tim Page's Nam* (New York, 1983), book jacket.
51 Martha Rosler, *Decoys and Disruptions: Selected Writings, 1975–2001* (Cambridge, MA, 2004), pp. 245–58.
52 Mark Reinhardt, Holly Edwards and Erina Duganne, eds, *Beautiful Suffering: Photography and the Traffic in Pain* (Chicago, IL, 2007).
53 Steve McCurry, *Portraits* (London, 1999); *National Geographic: 100 Best Pictures: Special Member's Edition* (2002); 'Steve McCurry: Poster', stevemccurry.com, accessed 12 September 2018.
54 The quotations come from the promotional text by Phaidon for the second edition of Steve McCurry, *Portraits* (London, 2013), www.phaidon.com, accessed 23 July 2018.
55 Holly Edwards, 'Cover to Cover: The Life Cycle of an Image in Contemporary Visual Culture', in *Beautiful Suffering*, ed. Reinhardt, Edwards and Duganne, pp. 75–92.
56 Simon James, 'Steve McCurry: Humanity in Colour', *RPS Journal*, CXLII/6 (2002), pp. 48–51, cited ibid., p. 79.
57 David Campbell, 'The Myth of Compassion Fatigue', in *The Violence of the Image: Photography and International Conflict*, ed. Liam Kennedy and Caitlin Patrick (London, 2014), pp. 97–124.
58 Brian Coe, 'The Rollfilm Revolution', in *The Story of Popular Photography*, ed. Colin Ford (Bradford, 1989), pp. 60–68; Nancy Martha West, *Kodak and the Lens of Nostalgia* (Charlottesville, VA, 2000), pp. 1–2, 41.
59 Michael Pritchard, *The History of Photography in 50 Cameras* (London, 2014), pp. 74–9.
60 Advertisement by camera retailer Walter Suckling Ltd, Auckland, *Auckland Star*, 20 September 1915, p. 10.
61 Hilary Roberts, 'Photography', www.1914-1918-online (2014).
62 Ibid.; Eberhard Demm, 'Censorship', www.1914-1918-online (2017).
63 Hilary Roberts, 'Photography'.
64 Janina Struk, *Private Pictures: Soldiers' Inside View of War* (London, 2011), pp. 28–9.
65 Diane Atkinson, *Mairi and Elsie Go to War: Two Extraordinary Women at the Western Front* (Cambridge, 2010).
66 Pippa Oldfield, 'Snapshots from No Man's Land', www.britishartstudies.ac.uk (2018).
67 G. Thomas, 'Indian Mutiny Veterans: The Tytlers', *History of Photography*, IX/4 (1985), pp. 267–73.
68 Cited in a report of a meeting of the Photographic Society of Bengal, published in *The Englishman*, 31 March 1859, cited in Roger Taylor and Larry J. Schaaf, *Impressed by Light: British Photographs from Paper Negatives, 1840–1860* (New York, 2007), p. 386.
69 Patricia Vettel-Becker, *Shooting from the Hip: Photography, Masculinity, and Postwar America* (Minneapolis, MN, 2005), pp. 37–42.
70 Cited in Scherman, 'Introduction', p. 8.
71 Richard Whelan, *Robert Capa: A Biography* (Lincoln, NE, 1994), pp. 80–83.
72 Irme Schaber, Richard Whelan and Kristen Lubben, eds, *Gerda Taro: From the Collection of the International Center of Photography* (New York, 2007); Cynthia Young, ed., *The Mexican Suitcase: The Rediscovered Spanish Civil War Negatives of Capa, Chim, and Taro* (New York and Göttingen, 2010).
73 Lucy Davies and Susan Meiselas, 'Blending In', *British Journal of Photography Special Issue: Women in Photography*, 4 (2014), p. 45.
74 Alistair Murphy and Elizabeth Elmore, *Fishermen and Kings: The Photography of Olive Edis* (Norwich, 2016).
75 Alberto del Pino, 'Relaciones entre fotografía y demás iconografía de la Guerra del Paraguay', *Folia Histórica del Nordeste*, 25 (2016), pp. 133–58; Miguel Angel Cuarterolo, 'Images of War', in *I Die with My Country: Perspectives on the Paraguayan War, 1864–1870*, ed. Hendrik Kraay and Thomas Whigham (London and Lincoln, NE, 2004), pp. 154–78.
76 Gary D. Sampson, 'Military Photography', in *Encyclopedia of Nineteenth-century Photography*, ed. John Hannavy (New York and London, 2008), p. 929.
77 Cécile Bishop, 'Photography, Race and Invisibility', *Photographies*, XI/2–3 (2018), pp. 193–213.
78 Jeffrey T. Sammons and John H. Morrow Jr, *Harlem's Rattlers and the Great War: The Undaunted 369th Regiment and the African American Quest for Equality* (Lawrence, KS, 2014).
79 John Tulloch and R. Warwick Blood, 'Iconic Photojournalism and Absent Images: Democratization and Memories of Terror', in *The Routledge Companion to News and Journalism*, ed. Stuart Allan (Abingdon, 2010), pp. 507–19.
80 Cited ibid., p. 514.

two: Military Vision

1 '20 Million Photos a Month', *Science News-letter*, 19 May 1945, pp. 315–17.
2 Eugène Durieu, 'Rapport', *Bulletin de la Société Française de Photographie Paris* (February 1856), p. 50. Cited in Mary Warner Marien, *Photography: A Cultural History*, 3rd edn (London, 2010), p. 100.

3 Steve Edwards, *The Making of English Photography: Allegories* (Philadelphia, PA, 2006), pp. 1–2.
4 Mark Haworth-Booth, *Camille Silvy: River Scene, France* (Malibu, CA, 1992), pp. 70–88.
5 Graham Saxby, 'Photogrammetry', in *The Oxford Companion to the Photograph*, ed. Robin Lenman (Oxford, 2005), p. 480; David Mattison, 'Aerial Photography', in *Encyclopedia of Nineteenth-century Photography*, ed. John Hannavy (New York and London, 2008), pp. 12–15.
6 Captain R. E. Donnelly, 'Photography, and its Application to Military Purposes', *Royal United Services Institution Journal*, V/XVI (1861), pp. 1–18.
7 Paul Becherer and Bill Woodburn, 'Photographs by the Photograph School of the Bengal Sappers and Miners: Photographs Taken in Afghanistan during the Campaigns of 1878–79 and 1879–80', www.phototheca-afghanica.ch, accessed 7 July 2018.
8 Michel Foucault, 'The Eye of Power: A Conversation with Jean-Pierre Barou and Michelle Perrot', in *Michel Foucault: Power/Knowledge: Selected Interviews and Other Writings*, ed. Colin Gordon (London, 1980), pp. 146–65.
9 Pat Hodgson, *Early War Photographs* (Reading, 1974), p. 20.
10 R. K. Anderson, 'History of the U.S. Army Visual Information Center, 1918–2002', www.armysignalocs.com, accessed 7 July 2018.
11 Daniel M. Davis, 'Russell, Andrew Joseph (1832–1909): American Photographer', in *Encyclopedia of Nineteenth-century Photography*, ed. Hannavy, pp. 1225–7.
12 Alan Trachtenberg, 'Albums of War: On Reading Civil War Photographs', *Representations*, 9 (1985), pp. 7, 25–7; 'African American Labourers on a Railroad', www.oah.org, 2018.
13 'History of Mapping the U.S. Civil War', www.loc.gov, accessed 7 July 2018.
14 For an image showing maps being copied in the field, see Jeff Rosenheim, *Photography and the American Civil War* (New York, 2013), p. 14.
15 Gary D. Sampson, 'Military Photography' in *Encyclopedia of Nineteenth-century Photography*, ed. Hannavy, pp. 929–32.
16 Stephen Sheehi, *The Arab Imago: A Social History of Portrait Photography, 1860–1910* (Princeton, NJ, 2016), p. 164; John De St Jorre, 'Pioneer Photographer of the Holy Cities', *Saudi Aramco World*, L/1 (1999), pp. 36–47; Claude W. Sui, 'Early Photography of the Holy Sites of Islam in the Arabian Peninsula', in *The Indigenous Lens?: Early Photography in the Near and Middle East*, ed. Markus Ritter and Staci G. Schweiwiller (Berlin and Boston, MA, 2018), pp. 111–44.
17 Engin Özendes, *Photography in the Ottoman Empire, 1839–1919* (Istanbul, 1987); Michèle Hannoosh, 'Practices of Photography: Circulation and Mobility in the Nineteenth-century Mediterranean', *History of Photography*, XL/1 (2016), pp. 3–27.
18 Elahe Helbig, 'Geographies Traced and Histories Told: The Photographic Documentation of Land and People by 'Abdullah Mirza Qajar, 1880s to 1890s', in *The Indigenous Lens*, ed. Ritter and Scheiwiller, pp. 79–109.
19 Helbig, 'Geoegraphies Traced', pp. 95–6.
20 Melanie Osborne, 'Mapping for War and Peace: Ordnance Survey's Maps and the First World War', https://history.blog.gov.uk, 13 May 2014; Ute Scheider, 'Maps, Geography and the Representation of Space', encyclopedia.1914-1918-online.net, 13 December 2017.
21 Mattison, 'Aerial Photography', pp. 12–15.
22 Denis Cosgrove and William L. Fox, *Photography and Flight* (London, 2010), p. 26.
23 Matt Peitersen, 'America's First NGA? The Myth and Mystery of Civil War Aerial Photography', http://disturbedgeographer.com, 3 August 2015.
24 Anderson, 'History of the U.S. Army Visual Information Center'.
25 E. Lee, *To the Bitter End* (London, 1985), p. 35, cited in Matt Pietersen, 'America's First NGA?'
26 Frank E. Rietz, 'Alfred Maul: A Pioneer of Camera Rocket', *Acta Astronautica*, XXIV (1991), pp. 365–75.
27 Terrence Finnegan, *Shooting the Front: Allied Aerial Reconnaissance in the First World War* (Washington, DC, 2006); Dan Gettinger, 'The Ultimate Way of Seeing: Aerial Photography in WWI', http://dronecenter.bard.edu, 2014; Charles Dusch, 'Reconnaissance and Observation', https://encyclopedia.1914-1918-online.net, 2018.
28 R. Kingslake, 'Lenses for Aerial Photography', *Journal of the Optical Society of America*, XXXII/3 (1942), pp. 129–34.
29 Cited in Roy M. Stanley II, *World War II Photo Intelligence* (New York, 1981), p. 16.
30 George Goddard, *Overview: A Lifelong Adventure in Aerial Photography* (New York, 1969), pp. 340–41; Jeremy Harwood, *World War II from Above: An Aerial View of the Global Conflict* (Minneapolis, MN, 2014), p. 50.
31 Christine Halsall, *Women of Intelligence: Winning the Second World War with Air Photos* (Stroud, 2012), pp. 158–66.
32 Ibid., p. 22.
33 The term was introduced in 1960 by Evelyn L. Pruitt of the U.S. Office of Naval Research. Paul R. Baumann, 'History of Remote Sensing, Aerial Photography', www.ononta.edu, 2014.
34 'The Cuban Missile Crisis of 1962: Presenting the Photographic Evidence Abroad', www.cia.gov, 19 March 2007; 'The Cuban Missile Crisis, 1962: The Photographs', http://nsarchive2.gwu.edu, accessed 7 July 2018.
35 Asif Siddiqi and Dwayne A. Day, 'Lies, Damned Lies, and Cover Stories: New Revelations About Satellite Secrecy During the Cold War', www.thespacereview.com, 20 November 2006; John Cloud, 'American Cartographic Transformations during the Cold War',

Cartography and Geographic Information Science, XXIX/3 (2002), pp. 261–82.
36 Anderson, 'U.S. Army Visual Information Center'.
37 Ibid.
38 Cited in Mark Dorrian, 'Drone Semiosis: Weaponry and Witnessing', *Cabinet*, 54 (2014), p. 51.
39 Lawrence D. Freedman, 'The Drone Revolution: Less Than Meets the Eye', *Foreign Affairs*, XCV/6 (2016), p. 158.
40 Dorrian, 'Drone Semiosis', pp. 52–4; Derek Gregory, 'From a View to a Kill: Drones and Late Modern War', *Theory, Culture and Society*, XXVIII/7–8 (2011), pp. 188–215.
41 Kitty Hauser, *Shadow Sites: Photography, Archaeology and the British Landscape, 1927–1955* (Oxford, 2007), pp. 164–7; Jananne Al-Ani and Nat Muller, 'Technologies of History', *Ibraaz*, 7 (2014), www.ibraaz.org.
42 Cited in Emily Pethick, 'Jananne Al-Ani: The Visit', www.luxonline.org.uk, 10 April 2005.
43 Jananne Al-Ani and Nat Muller, 'Technologies of History', www.ibraaz.org, accessed 7 July 2018.
44 Wilco Versteeg, 'War Beyond Photography: Digitally Embellished Imaging and Contemporary Conflicts', *Arts of War and Peace Review*, II/1 (May 2014), p. 43; Pippa Oldfield, 'Nothing to See? The Future of War Photography Symposium, Imperial War Museum North', *Media, War and Conflict*, III/3 (2010), pp. 369–71.
45 The contributors to the collective project are Ali Rez, Saks Afridi, Assam Khalid, Akash Goel, Insiya Syed, Noor Behram, Jamil Akhtar and the JR InsideOut project.
46 Dorrian, 'Drone Semiosis', pp. 52–5.
47 The group featured on CNN, *The Guardian, The Independent* and other news sites. https://notabugsplat.com, accessed 7 July 2018.
48 Trevor Paglen, 'The Other Night Sky', www.paglen.com, accessed 7 July 2018.
49 Bryan Gardiner, 'Photographer Documents Secret Satellites – All 189 of Them', www.wired.com, 21 June 2008; David Markus, 'Trevor Paglen', www.artinamerica.com, 19 October 2017.
50 Patricia A. Nelson, 'Competition and the Politics of War: The Global Photography Industry, *c.* 1910–60', *Journal of War and Culture Studies: The Business of War Photography*, IX/2 (2016), pp. 115–32.
51 Madeleine Hill Vedel, Mariko Takeuchi and Osamu Hiraki, 'Japan', in *The Oxford Companion to the Photograph*, ed. Robin Lenman (Oxford, 2005), pp. 327–31.
52 Nelson, 'Competition and the Politics of War', p. 120.
53 Rachel Snow, 'Photography's Second Front: Kodak's "Serving Human Progress Through Photography" Institutional Advertising Campaign', *Journal of War and Culture Studies: The Business of War Photography*, IX/2 (2016), pp. 151–81.
54 O.F.G. Hogg, *Artillery: Its Origin, Heyday and Decline* (London, 1970), pp. 174–6.
55 Ned O'Gorman and Kevin Hamilton, 'EG&G and the Deep Media of Timing, Firing, and Exposing', *Journal of War and Culture Studies: The Business of War Photography*, IX/2 (2016), pp. 189–91; 'Harold "Doc" Edgerton: Visionary Engineer', http://edgerton-digital-collections.org, accessed 7 July 2018.
56 O'Gorman and Hamilton, 'EG&G', p. 190.
57 Michael Peres, *Laboratory Imaging and Photography: Best Practices for Photomicrography and More* (New York, 2017), p. 20; Roger W. Hicks, 'Infrared Photography', in *The Oxford Companion to the Photograph*, ed. Robin Lenman (Oxford, 2005), pp. 309–10.
58 Peres, *Laboratory Imaging and Photography*, p. 21.
59 Armand Mattelart, *The Globalization of Surveillance* (Cambridge, 2010), p. 99, p. 112; René De La Pedraja, *Wars of Latin America, 1948–1982: The Rise of the Guerrillas* (Jefferson, NC, 2013), p. 169.
60 'UNODC and Illicit Crop Monitoring', www.unodc.org, accessed 14 September 2018.
61 'Decode Darfur', decorders.amnesty.org, accessed 14 September 2018.
62 Wilco Versteeg, 'War Beyond Photography: Digitally Embellished Imaging and Contemporary Conflicts', *Arts of War and Peace Review*, II/1 (May 2014), pp. 40–41.
63 William White, 'The Microdot: Then and Now', *International Journal of Intelligence and Counter Intelligence*, III/2, pp. 249–69.
64 Geoffrey Belknap, *From a Photograph: Authenticity, Science and the Periodical Press, 1870–1890* (London, 2016).
65 Frederic Luther, 'René Dagron, Inventor of Microfilm', *History of Photography*, XX/IV (1996), pp. 345–61.
66 'Morale: Scarlet Scourge', *Army and Navy*, 14 February 1944; David A. Norris, 'V-Mail: The GIs' Lifeline', *America in WWII Special Issue: GIs in World War II* (Autumn 2012), pp. 52–5; John Bush Jones, *All-out for Victory! Magazine Advertising and the World War II Home Front* (Lebanon, NH, 2009), pp. 174–6.
67 Oliver Wendell Holmes, 'Photographic Phases', *New York Times*, 21 July 1862.
68 Klaus Tenfelde, *Pictures of Krupp: Photography and History in the Industrial Age* (London, 2005).
69 Deac Russell, 'Ernst Mach (1838–1916): Moravian Physicist', in *Encyclopedia of Nineteenth-century Photography*, ed. Hannavy, pp. 879–80.
70 Deac Rossell, 'Ottomar Anschütz (1846–1907): German Photographer and Moving Picture Pioneer', in *Encyclopedia of Nineteenth-century Photography*, ed. Hannavy, pp. 46–8.
71 Cited in David E. Scherman, 'Introduction: The Photographer Under Fire', in *Life at War* (New York, 1981), p. 6.
72 O'Gorman and Hamilton, 'EG&G', pp. 182–210.
73 Olivier Debroise, *Mexican Suite: A History of Photography in Mexico*, trans. and rev. by Olivier Debroise and Stella de Sá Rego (Austin, TX, 2001), pp. 164–7.

74 It is not known if Bontecou took the photographs himself, or commissioned others to make them. Shirley Samuels, *Facing America: Iconography and the Civil War* (New York, 2004), pp. 89–93; Stanley B. Burns, *Shooting Soldiers: Civil War Medical Photography by R. B. Bontecou* (New York, 2011).

75 Jeff Rosenheim, *Photography and the American Civil War* (New York, 2013), p. 178.

76 Adrian M. K. Thomas, 'The First 50 Years of Military Radiology, 1895–1945', *European Journal of Radiology*, 63 (2007), pp. 214–19.

77 Beverly A. Spirit and Patricia A. Randall, 'Radiologic History Exhibit: The Role of Women in Wartime Radiology', *Radiographics*, xv/3 (1995), pp. 641–52; Peter E. Palmquist, *Elizabeth Fleischmann: Pioneer X-Ray Photographer* (Berkeley, CA, 1990); 'Mauser Bullet Lies Deep in His Brain', *San Francisco Call*, 28 September 1899, p. 5; Percy Brown, 'Elizabeth Fleischmann-Anscheim (1859–1905)', *American Martyrs to Science Through the Roentgen Rays* (Springfield, IL, 936), reprinted in *American Journal of Roentgenology*, 164 (1995), pp. 497–9.

78 Thomas, 'The First 50 Years of Military Radiology', pp. 217–18.

79 A. F. Wallace, 'The Early History of Clinical Photography for Burns, Plastic and Reconstructive Surgery', *British Journal of Plastic Surgery*, XXXVIII (1985), pp. 451–65.

80 'Bravemind: Virtual Reality Exposure Therapy', http://ict.usc.edu, accessed 1 February 2018.

81 Lisa Barnard, *Hyenas of the Battlefield, Machines in the Garden* (London, 2014), pp. 19–90.

82 Lisa Barnard, 'Post Traumatic Stress Disorder PTSD', http://lisabarnard.co.uk, accessed 10 February 2019.

three: Home Fronts

1 Gail Buckland, 'Nicholls, Horace Walter, 1867–1941', in *Encyclopedia of Nineteenth-century Photography*, ed. John Hannavy (New York and London, 2008), pp. 1000–1001.

2 Angela Woolacott, *On Her Their Lives Depend: Munitions Workers in the Great War* (Berkeley, CA, 1994).

3 Colin Harding, 'From Desk to Trench', www.blogs.iwm.org.uk, 14 September 2017.

4 Jennifer Wellington, *Exhibiting War: The Great War, Museums, and Memory in Britain, Canada, and Australia* (Cambridge, 2017), p. 74. A photograph of the 'shrine' may be viewed online at www.iwm.org using the image reference Q31113.

5 Matthew Shaul, ed., *Beyond the Battlefields: Käthe Buchler's Photographs of Germany in the Great War* (Hatfield, 2018).

6 Samuel F. Villela, *Sara Castrejón: Fotógrafa de la Revolución* (Mexico City, 2010).

7 Paul J. Vanderwood and Frank N. Samponaro, *Border Fury: A Picture Postcard Record of Mexico's Revolution and U.S. War Preparedness, 1910–1917* (Albuquerque, NM, 1988).

8 Villela, *Sara Castrejón*, p. 57; Elizabeth Salas, *Soldaderas in the Mexican Military: Myth and History* (Austin, TX, 1990).

9 Stephen Badsey, 'Propaganda: Media in War Politics', www.1914-1918-online.net, 2014.

10 'Will These Flags be Raised Against a Foreign Power?', *PM*, 1 August 1940.

11 David C. Earhart, 'Kamikazefication and Japan's Wartime Ideology', *Critical Asian Studies*, XXXVII/4 (2005), pp. 569–96

12 The image was originally published in David C. Earhart, *Certain Victory: Images of World War II in the Japanese Media* (London and New York, 2008), p. 409.

13 Ibid., p. 587.

14 Bodo von Dewitz and Robert Lebeck, eds, *Kiosk: A History of Photojournalism, 1839–1973* (Cologne, 2001), pp. 17, 192, 213–19; Jordan Henry, '"We Europeans": Signal Magazine and Political Collaboration in German-occupied Europe, 1940–1945', BA thesis (Ohio State University, 2017).

15 'U.S. is Losing the War of Words', *Life*, 22 March 1943, p. 11.

16 Claude Bellanger, *Histoire generale de la presse francaise, tome 4: de 1940 a 1958* (Paris, 1975), p. 12, cited in Henry, '"We Europeans"', p. 15.

17 Henry, '"We Europeans"', p. 14.

18 Janina Struk, *Private Pictures: Soldiers' Inside View of War* (London, 2011), pp. 44–5.

19 Nancy West, *Kodak and the Lens of Nostalgia* (Charlottesville, VA, and London, 2000), pp. 166–99.

20 Pippa Oldfield, 'Calling the Shots: Women's Photographic Engagement with War in Hemispheric America, 1910–1990', PhD thesis, University of Durham (2016), pp. 162–73; John Bush Jones, *All-out for Victory! Magazine Advertising and the World War II Home Front* (Lebanon, NH, 2009), pp. 160–72.

21 Vron Ware, *Military Migrants: Fighting for YOUR Country* (Basingstoke, 2012), pp. 39–40.

22 Jill Cole, 'Birds' and 'Paintball Portraits', www.jillcole.com, accessed 1 February 2018.

23 'Don McCullin, Northern Ireland, Londonderry', www.nationalgalleries.org, accessed 15 September 2018.

24 The *Sunday Times Magazine*, 19 December 1971.

25 Cited in Barbie Zelizer, 'Finding Aids to the Past: Bearing Personal Witness to Traumatic Public Events', *Media, Culture, and Society*, XXIV/5 (2002), p. 12.

26 'Updates', www.hereisnewyork.org, accessed 1 February 2018.

27 *L'Illustration*, 1 July 1848, cited in Dewitz and Lebeck, *Kiosk*, pp. 22–6.

28 Éric Germain, 'Out of Sight, Out of Reach: Moral Issues in the Globalization of the Battlefield', *International Review of the Red Cross*, XCVII/900 (2015), pp. 1065–97.
29 'Germans Attack "the Englishman's Home" at Last', *The War Illustrated*, 26 December 1914.
30 The image was first published in *Regards*, 14 May 1936. Erika Zerwes, 'The Mother From Extremadura and the Idea of a Photographic Icon', *Roots and Routes: Research on Visual Cultures*, V/17 (2015).
31 The authorship of the cover is not known. Horacio Fernández, ed., *Photobooks: Spain, 1905–1977* (Madrid, 2014), pp. 92–9.
32 'Thérèse Bonney on Eve of Exhibition', press release issued by the Museum of Modern Art, New York, 5 December 1940, www.moma.org, accessed 30 June 2018.
33 Heide Fehrenbach, 'Children and Other Civilians: Photography and the Politics of Humanitarian Image-making', in *Humanitarian Photography: A History*, ed. Heide Fehrenbach and Davide Rodogno (Cambridge, 2015), pp. 186–8; Lesley Blanch, 'History in the Taking: Thérèse Bonney, War Photographer', U.S. *Vogue*, 1 July 1943; Nan Robertson, 'In a Life of Firsts, She Has Few Regrets', *New York Times*, 25 July 1976; Hanna Diamond, 'Representing Defeat: Photographic Images of the French Exodus of 1940', *Journal of War and Culture Studies*, I/3 (208), pp. 275–92.
34 Francesco D'Orazio, 'Journey of an Image: From a Beach in Bodrum to Twenty Million Screens Across the World', in *The Iconic Image on Social Media: A Rapid Research Response to the Death of Aylan Kurdi*, ed. F. Vis and O. Goriunova (Sheffield, 2015), pp. 11–18.
35 Heidi Fehrenbach and Davide Rodogno, '"A Horrific Photo of a Drowned Syrian Child": Humanitarian Photography and NGO Media Strategies in Historical Perspective', *International Review of the Red Cross*, XCVII/900 (2015), p. 146.
36 Sarah C. Nelson, 'Katie Hopkins Silent On "Cockroach Migrants" as *The Sun* Launches Crisis Campaign For Aylan', www.huffingtonpost.co.uk, 2015.
37 Lucy Mayblin, 'Politics, Publics, and Aylan Kurdi', in *The Iconic Image on Social Media*, ed. Vis and Goriunova, pp. 42–3.
38 Sabine Kriebel, 'Photomontage in the Year 1932: John Heartfield and the National Socialists', *Oxford Art Journal*, XXXI/1 (2008), pp. 99–127.
39 In 2004 Martha Rosler rebooted her series *Bringing the War Home*, turning her critical eye to U.S. intervention in Iraq and Afghanistan. See Katherine Brickell, 'Geopolitics of Home', *Geography Compass*, VI/10 (2012), pp. 575–88.
40 Ed Barber, *Peace Moves: Nuclear Protest in the 1980s* (London, 1984); Helen Trompeteler, 'Format Photographers Agency, 1983–2003', information sheet for exhibition at the National Portrait Gallery (London, 2010); Lyn Smith, *People Power* (London, 2017), pp. 156–85.
41 The collage in fact fused two elements: Blair photographing himself with naval cadets during his election campaign of 2005, and a dramatic oil explosion. See www.kennardphillipps.com, accessed 30 June 2018.
42 Communication with the artists, June 2018; Jonathan Jones, 'The Tony Blair "Selfie" Photo Op Will Have a Place in History', *The Guardian*, 15 October 2013.

four: Secrets and Exposures

1 Benjamin originally posed his question in the article 'Small History of Photography', first published in 1931 and reproduced in Walter Benjamin, *On Photography*, trans. Esther Leslie (London, 2015).
2 Stacy Hand, 'Microphotography', in *Encyclopedia of Nineteenth-century Photography*, ed. John Hannavy (New York and London, 2008), pp. 925–9.
3 William White, 'The Microdot: Then and Now', *International Journal of Intelligence and Counter Intelligence*, III/2 (1989), p. 254.
4 Ibid., pp. 249–69.
5 Enn Hendre, 'Fifty Years of the Sub-miniature Minox', *History of Photography*, XIII/1 (1989), pp. 9–17.
6 Michael Pritchard, *Spy Camera: A Century of Detective and Subminiature Cameras* (London, 1993); Roger W. Hicks, 'Sub-miniature Photography', in *The Oxford Companion to the Photograph*, ed. Robin Lenman (Oxford, 2005), pp. 610–11; Robin Lenman, 'Zapp, Walter', in *The Oxford Companion to the Photograph*, ed. Lenman, p. 697.
7 Felix Müller, Curator, Stasimuseum, Berlin, email correspondence with the author, 4 September 2018.
8 'Tasks of the BSTU', www.BSTU.bund.de, accessed 1 February 2018.
9 Simon Menner, *Top Secret: Images from the Stasi Archives* (Berlin, 2013).
10 'Historia', Archivo Histórico de la Policía Nacional, archivohistoricopn.org, accessed 14 September 2018; Kirsten Weld, *Paper Cadavers: The Archives of Dictatorship in Guatemala* (Durham, NC, 2014).
11 Hasan Elahi, 'You Want to Track Me? Here You Go, FBI', *New York Times*, 30 October 2011. A live feed of the *Tracking Transience* project can be viewed at http://elahi.umd.edu.
12 Edhem Eldem, 'Powerful Images: The Dissemination and Impact of Photography in the Ottoman Empire, 1870–1914', in *Camera Ottomana: Photography and Modernity in the Ottoman Empire, 1840–1914*, ed. Zeynep Çelik and Edhem Eldem (Istanbul, 2015), pp. 106–53.
13 'The Events of 1903 in Macedonia as Presented in European Diplomatic Correspondence', www.macedonian-heritage.gr, 2018.
14 Cited in Eldem, 'Powerful Images', p. 121.

15 Bodo von Dewitz, *'So wird bei uns der Krieg geführt': Amateurfotografie im Ersten Weltkrieg* (Munich, 1989); Janina Struk, *Private Pictures: Soldiers' Inside View of War* (London, 2011); Petra Bopp, *Fremde im Visier: Fotoalben aus dem Zweiten Weltkrieg* (Bielefeld, 2012); Frances Guerin, *Through Amateur Eyes: Film and Photography in Nazi Germany* (Minneapolis, MN, 2012).
16 Struk, *Private Pictures*, pp. 89–108.
17 Cited ibid., p. 95.
18 Alison M. Moore, 'History, Memory and Trauma in Photography of the Tondues: Visuality of the Vichy Past through the Silent Image of Women', *Gender and History*, XVII/3 (2005), pp. 657–81.
19 *La Grande délivrance de Paris* was held at Musée Carnavalet, Paris, from November 1944 to January 1945; the catalogue was published in 1945. Charles Chadwyck-Healey, 'The Literature of the Liberation 1944–1946', *The Book Collector* (March 2015).
20 Antonio M. Taguba, 'U.S. Army Article 15-6 Investigation of the 800th Military Police Brigade' (Washington, DC, May 2004), pp. 16–17.
21 Philip Gourevitch, 'Exposure: The Woman Behind the Camera at Abu Ghraib', *New Yorker*, 24 March 2008.
22 Seymour M. Hersh, 'Torture at Abu Ghraib', *New Yorker*, 10 May 2004; Anderson Cooper, 'Exposing the Truth About Abu Ghraib', www.cbsnews.com, 7 December 2006.
23 Rachel Hughes, 'The Abject Artefacts of Memory: Photographs from Cambodia's Genocide', *Media, Culture and Society*, XXV (2003), pp. 23–44.
24 Seth Mydens, 'Out from Behind a Camera at a Khmer Torture House', www.nytimes.com, 26 October 2007.
25 Nhem Ein stated that the label was laminated in plastic and stuck to the boy's damp skin, rather than being pinned. Cited in Rachel Hughes, 'The Abject Artefacts of Memory: Photographs From Cambodia's Genocide', *Media, Culture and Society*, XXV (2003), p. 40.
26 Marianne Hirsch, *Family Frames: Photography, Narrative and Postmemory* (Cambridge, MA, 1997), p. 267.
27 Conservation has been due, in part, to U.S. photojournalists Douglas Niven and Christopher Riley, later known as the Photo Archive Group. Eamonn Carrabine, 'Iconic Power, Dark Tourism and the Spectacle of Suffering', in *The Palgrave Handbook of Prison Tourism*, ed. Jacqueline Z. Wilson, Sarah Hodgkinson, Justin Piché and Kevin Walby (London, 2017), pp. 12–36.
28 The phrase 'messengers of peace' comes from the museum's mission statement. 'Tuol Sleng Genocide Museum', http://tuolsleng.gov.kh, accessed 1 February 2018.
29 Jasmine Alinder, *Moving Images: Photography and the Japanese American Incarceration* (Chicago and Urbana, IL, 2009); Wendy Kozol, 'Relocating Citizenship in Photographs of Japanese Americans during World War II', in *Haunting Violations: Feminist Criticism and the Crisis of the 'Real'*, ed. Wendy S. Hesford and Wendy Kozol (Chicago and Urbana, IL, 2001), pp. 217–50.
30 Linda Gordon and Gary Y. Okihiro, eds, *Impounded: Dorothea Lange and the Censored Images of Japanese American Internment* (New York, 2006). The images have now been digitized by the National Archives and are available to view online at www.archives.gov.
31 Sharon Yamato, 'Tōyō Miyatake', Densho Encyclopedia, https://encyclopedia.densho.org, accessed 30 June 2018.
32 Gina L. Wenger, 'Documentary Photography: Three Photographers' Standpoints on the Japanese-American Internment', *Art Education*, LX/5 (2007), pp. 33–8.
33 Karin Higa, 'Tōyō Miyatake and "Our World"', in *Only Skin Deep: Changing Visions of the American Self*, ed. Coco Fusco and Brian Wallis (New York, 2003); 'JARDA: Japanese American Relocation Digital Archive', www.calisphere.org, accessed 30 June 2018.
34 Natalia Fortuny, 'Imágenes sobrevivientes: Fotografía y memoria en una obra de Paula Luttringer', *Clepsidra. Revista Interdisciplinaria de Estudios sobre Memoria*, 1 (2014), pp. 14–27.
35 Christl Verduyn and Conny Steenman Marcusse, 'Emmy Andriesse, Dutch Wartime Photographer: The Hunger Winter of 1945', in *Bearing Witness: Perspectives on War and Peace from the Arts and Humanities*, ed. Sherrill Grace, Patrick Imbert and Tiffany Johnstone (Montreal, 2012), pp. 121–44.
36 *Amsterdam Tijdens den Hongerwinter*, with an introduction by Max Nord and photographs by Emmy Andriesse, Cas Oorthuys, Ad Windig, Krijn Taconis and others (Amsterdam, 1947).
37 Georges Didi-Huberman, *Images in Spite of All: Four Photographs from Auschwitz* (Chicago, IL, 2008); Dan Stone, *History, Memory and Mass Atrocity* (Portland, OR, 2006), p. 16; and Janina Struk, *Photographing the Holocaust: Interpretations of the Evidence* (London, 2004), pp. 112–15.
38 Jean-Claude Pressac, *Auschwitz: Technique and Operation of the Gas Chambers* (New York, 1989), p. 422.
39 Kerri Macdonald, 'A Fighter with a Camera in Apartheid Era South Africa', https://lens.blogs.nytimes.com, 21 September 2012.
40 C. Gerald Fraser, 'A Black Photojournalist Depicts Apartheid Life', *New York Times*, 14 May 1978, p. 48; 'Peter Sexford Magubane', www.sahistory.org.za, 17 February 2011.
41 Kathleen Collins, 'Living Skeletons: Carte-de-visite Propaganda in the American Civil War', *History of Photography*, XII/2 (1988), pp. 103–20.
42 Judith A. Giesberg, '"Eye of History": Looking at Civil War Prisoners of War', in *Lens of War: Exploring Iconic Photographs of the Civil War*, ed. J. Matthew Gallman and Gary W. Gallagher (Athens, GA, 2015), p. 187.
43 Collins, 'Living Skeletons', pp. 114–15.
44 Barbie Zelizer, *Remembering to Forget: Holocaust Memory through the Camera's Eye* (Chicago, IL, 1998), p. 146; Sharon Sliwinski,

'Visual Testimony: Lee Miller's Dachau', *Journal of Visual Culture*, IX (2010), p. 389.

45 Sliwinski, 'Visual Testimony', p. 402; Zelizer, *Remembering to Forget*, pp. 144–6; and Struk, *Photographing the Holocaust*, pp. 141–3.

46 Pippa Oldfield, 'Calling the Shots: Women's Photographic Engagement with War in Hemispheric America, 1910–1990', PhD thesis, Durham University, 2016, http://etheses.dur.ac.uk, pp. 141–62.

47 R. Jenkins, ed., *Nagasaki Journey: The Photographs of Yōsuke Yamahata: August 10, 1945* (Tokyo, 1995).

48 Barbara Marcón, 'Hiroshima and Nagasaki in the Eye of the Camera', *Third Text*, XXV/6 (2011), pp. 787–9.

49 Elizabeth D. Heineman, ed., *Sexual Violence in Conflict Zones: From the Ancient World to the Era of Human Rights* (Philadelphia, PA, 2011), pp. 1–14.

50 Peter Kort Zegers and Douglas W. Druick, *Windows on the War: Soviet TASS Posters at Home and Abroad, 1941–1945* (Chicago, IL, 2011).

51 Valerie Wieskamp, 'Sexual Assault and the Mai Lai Massacre: The Erasure of Sexual Violence from Public Memory of the Vietnam War', in *Mythologizing the Vietnam War: Visual Culture and Mediated Memory*, ed. Jennifer Good, Paul Lowe, Brigitte Lardinois and Val Williams (Newcastle upon Tyne, 2014), pp. 127–43.

52 Haeberle's photographs were first published in Joseph Eszterhas, 'Camerman Saw GIs Slay 100 Villagers', *Plain Dealer*, 20 November 1969; and subsequently as a colour feature, 'The Massacre at Mylai', *Life*, 5 December 1969, pp. 36–45.

53 Wieskamp, 'Sexual Assault', p. 129.

54 Seivan Salim, 'Escaped', www.mapofdisplacement.com, accessed 1 February 2018; Seivan M. Salim, 'The Women Who Escaped ISIS', www.thedailybeast.com, 2015.

55 Michael Godbey, 'Confronting Horror: Emily Hobhouse and the Concentration Camp Photographs of the South African War', *Kronos*, XXXII (2006), pp. 34–48.

56 Ibid., pp. 42–4.

57 Emily Hobhouse, *The Brunt of War and Where it Fell* (London, 1902), p. 215.

58 Ana Carden-Coyne, *Reconstructing the Body: Classicism, Modernism, and the First World War* (Oxford, 2009), pp. 157–8.

59 Ibid., pp. 110–59.

60 Susanne Brandt, 'Friedrich, Ernst', encyclopedia.1914-1918-online.net, 2015.

61 Brandt, 'Friedrich, Ernst', encyclopedia.1914-1918-online.net.

62 Ernst Friedrich, *Krieg dem Kriege* (Berlin, *c.* 1930), pp. 235–9.

63 Ana Longoni, 'Fotos y siluetas: dos estrategias en la representación de los desaparecidos', in *Los desaparecidos en la Argentina: Memorias, representaciones e ideas (1983–2008)*, ed. E. Crenzel (Buenos Aires, 2010), pp. 35–57.

64 Cora Gamarnik, 'La construcción de la imagen de las Madres de Plaza de Mayo a través de la fotografía de prensa', *Afuera: Estudios de crítica cultural*, 9 (2010), www.revistaafuera.com.

65 Matilde Mellibovsky, *Circle of Love over Death: Testimonies of the Mothers of the Plaza de Mayo*, trans. Maria Proser and Matthew Proser (Willimantic, CT, 1997).

66 Laurel Reuter, *Los Desaparecidos / The Disappeared* (Grand Forks, ND, 2006).

67 The original Madres de Plaza de Mayo group split in 1986. Fernando J. Bosco, 'Human Rights Politics and Scaled Performances of Memory: Conflicts Among the Madres de Plaza de Mayo in Argentina', *Social and Cultural Geography*, 5 (2004), pp. 381–402.

68 Leonardo Mellibovsky, email communication with the author, July 2018; Pippa Oldfield, 'Exploded Albums: Re-working the Family Photo Album in Women's Protests Against State Terrorism in Argentina', in *The Photograph and the Album: Histories, Practices, Futures*, ed. Jonathan Carson, Rosie Miller and Theresa Wilkie (Edinburgh and Boston, MA, 2013), pp. 121–45.

69 Aisling Walsh, 'Impunity Reigns: Threats to the Historical Archive of the National Police in Guatemala', *Democracia Abierta*, www.opendemocracy.net, 2 September 2018; 'APHN: Digital Archive of the Guatemalan National Police Historical Archive', University of Texas Libraries, ahpn.lib.utexas/edu, accessed 14 September 2018.

70 'Assad's Killing Machine Exposed: Implications for U.S. Policy', House Foreign Affairs Committee, held 31 July 2014, Capitol Hill, Washington, DC.

71 Committee on Foreign Affairs, 'HR 1677: Caesar Syria Civilian Protection Act of 2017', https://foreignaffairs.house.gov, 2017.

five: Legacies

1 Jeanne Moutoussamy-Ashe, *Viewfinders: Black Women Photographers* (New York, 1986), pp. 34–9.

2 Laura Augusta Lindenberger Wellen, 'Looking Forward Together: Three Studies of Artistic Practice in the South, 1920–1940', PhD thesis, University of Texas at Austin (2012), pp. 111–19.

3 'Unidentified Girl in Mourning Dress Holding Framed Photograph of her Father as a Cavalryman with Sword and Hardee Hat', www.loc.gov, accessed 1 February 2018.

4 New Focus, eds, *No Man's Land: Young People Uncover Women's Viewpoints on the First World War* (Bradford, 2017), pp. 28–33.

5 Geoffrey Batchen, *Forget Me Not: Photography and Remembrance* (New York, 2004).

6 Geoffrey Batchen, email correspondence with the author, 3 July 2018.

7 Wendy Red Star, 'Medicine Crow and the 1880 Crow Peace Delegation', www.wendyredstar.com, accessed 1 February 2018; '1880 Crow Peace Delegation', http://portlandartmuseum.us, accessed 1 February 2018.

8 See www.memoriaabierta.org.ar, accessed 14 September 2018.

9 See tuolsleng.gov.kh, accessed 14 September 2018.

10 See galerija110795.ba, accessed 14 September 2018.

11 Pippa Oldfield, 'A Thrilling Day Out: Air Shows and Military Spectacle', in *The Home Front*, ed. Melanie Friend (Stockport, 2013), n.p.

12 A. V. Seaton, 'War and Thanatourism: Waterloo 1815–1914', *Annals of Tourism Research*, XXVI/1 (1999), pp. 130–58.

13 Kitty Hauser, *Shadow Sites: Photography, Archaeology and the British Landscape, 1927–1955* (Oxford, 2007), pp. 30–56.

14 Coombs died in 1991; subsequent revisions of the book were undertaken by the publisher, After the Battle.

15 Léa Eouzan, '*Histoire(s) contemporaine(s)*', www.leaeouzan.com, accessed 1 February 2018.

16 Hauser, *Shadow Sites*, pp. 57–104.

17 Olive Edis, 'The Record of a Journey to Photograph the British Women's Services Overseas. Begun on Sunday March 2nd 1919', Olive Edis Papers and Photographs, Misc. O. Edis 89/19/1, Imperial War Museum.

18 Keith F. Davis, *George N. Barnard: Photographer of Sherman's Campaign* (Kansas City, MO, 1990); Simon Norfolk, *Afghanistan: Chronotopia* (Stockport, 2002).

19 Sandra Bustamante, 'Fotografía y política: La serie de retratos de Juan Travnik sobre Malvinas', www.revistaojosrojos.com, accessed 30 September 2018.

20 Madeleine Hill Vedel, Mariko Takeuchi and Osamu Hiraki, 'Japan', in *The Oxford Companion to the Photograph*, ed. Robin Lenman (Oxford, 2005), p. 330.

21 Kyoko Iriye Selden and Mark Selden, eds, *The Atomic Bomb: Voices from Hiroshima and Nagasaki* (New York and London, 1989).

22 Janet Farrell Brodie, 'Radiation Secrecy and Censorship after Hiroshima and Nagasaki', *Journal of Social History*, XLVIII/4 (2015), pp. 842–64.

23 Philip Jones Griffiths, *Agent Orange: 'Collateral Damage' in Viet Nam* (London, 2003).

24 'The Aftermath Project: The Story', http://theaftermathproject.org, accessed 1 February 2018; Nathalie Herschdorfer, *Afterwards: Contemporary Photography Confronting the Past* (London, 2011); Simon Baker and Shoair Mavlian, eds, *Conflict, Time, Photography*, exh. cat., Tate, London (2014).

25 Kathleen Palmer, *Women War Artists* (London, 2011), pp. 64–5.

26 Roland Barthes, *Camera Lucida* [1981], trans. Richard Howard (London, 2000), pp. 80–81.

27 Paul Bonaventura, 'Foreword', *Shot at Dawn*, ed. Paul Bonaventura and Chloe Dewe Mathews (Madrid, 2014), n.p.

28 'Nicaragua', www.susanmeiselas.com, accessed 14 September 2018.

29 'Susan Meiselas: Nicaragua, Reframing History, 2004', www.spiritofmarckarlin.com, 2 July 2015.

30 Susan Meiselas, 'Return to Nicaragua: The Aftermath of Hope', in *Rites of Return: Diaspora Poetics and the Politics of Memory*, ed. Marianne Hirsch and Nancy K. Miller (New York, 2011), pp. 166–70.

31 Denise Chong, *The Girl in the Picture: The Remarkable Story of Vietnam's Most Famous Casualty* (London and New York, 2001).

32 Rosy Martin and Jo Spence, 'New Portraits for Old: The Use of the Camera in Therapy', *Feminist Review*, 19 (1985), pp. 66–92; David A. Krauss and Jerry L. Fryrear, *Phototherapy in Mental Health* (Springfield, IL, 1983).

33 Riley Sharbonno and Monica Haller, *Riley and His Story: Me and My Outrage, You and Us* (Paris, 2009).

34 Pete Brook, 'Veterans' Photo Scrapbooking Offers New Therapy, Window into Experiences', www.wired.com, 5 July 2012.

35 'The Books (Objects for Deployment)', www.veteransbookproject.com, accessed 14 September 2018.

36 Caroline Wang and Mary Ann Burris, 'Photovoice: Concept, Methodology, and Use for Participatory Needs Assessment', *Health Education and Behaviour*, XXIV/3 (1997), pp. 369–87.

37 Tiffany Fairey, 'Shooting Kabul', www.tiffanyfairey.co.uk, accessed 1 February 2018.

38 Tessa Bunney, 'The Women of UCT6', www.tessabunney.co.uk, accessed 8 August 2018; Tessa Bunney, '*The Women of UCT6*', *FT Weekend Magazine*, 6 April 2013.

39 Frank Möller, 'From Aftermath to Peace: Reflections on a Photography of Peace', *Global Society*, XXXI/3 (2017), pp. 315–35.

Select Bibliography

Alinder, Jasmine, *Moving Images: Photography and the Japanese American Incarceration* (Chicago and Urbana, IL, 2009)

Allbeson, Tom, and Pippa Oldfield, 'War, Photography, Business: New Critical Histories', *Journal of War and Culture Studies*, IX/2 (2016), pp. 94–114

Atkinson, Diane, *Mairi and Elsie Go to War: Two Extraordinary Women at the Western Front* (Cambridge, 2010)

Baker, Simon, and Shoair Mavlian, eds, *Conflict, Time, Photography* (London, 2014)

Baring, Louis, *Emmy Andriesse: Hidden Lens* (Amsterdam, 2013)

Barnard, Lisa, *Hyenas of the Battlefield, Machines in the Garden* (London, 2014)

Barton, Peter, and Jeremy Banning, *The Battlefields of the First World War: The Unseen Panoramas of the Western Front* (London, 2013)

Barthes, Roland, *Camera Lucida*, trans. Farrar, Straus and Giroux [1981] (London, 2000)

Batchen, Geoffrey, *Forget Me Not: Photography and Remembrance* (New York, 2004)

—, Mick Gidley, Nancy K. Miller and Jay Prosser, eds, *Picturing Atrocity: Photography in Crisis* (London, 2012)

Beckers, Marion, and Elisabeth Moortgat, eds, *Kriegsfotografinnen*, a special edition of *Fotogeschichte – Beiträge zur Geschichte und Ästhetik der Fotografie*, XXXIV/134 (2014)

Belknap, Geoffrey, *From a Photograph: Authenticity, Science and the Periodical Press, 1870–1890* (London, 2016)

Berch, Bettina, *The Woman Behind the Lens: The Life and Work of Frances Benjamin Johnston, 1864–1952* (Charlottesville, VA, 2000)

Berumen, Miguel, *1911: La batalla de Ciudad Juárez / II Las imágenes* (Mexico City, 2003)

Bishop, Cécile, 'Photography, Race and Invisibility', *Photographies*, XI/2–3 (2018), pp. 193–213

Bopp, Petra, *Fremde im Visier: Fotoalben aus dem Zweiten Weltkrieg* (Bielefeld, 2012)

Brothers, Caroline, *War and Photography: A Cultural History* (Abingdon, 1997)

Burns, Stanley B., *Shooting Soldiers: Civil War Medical Photography by R. B. Bontecou* (New York, 2011)

Calvin, Paula E., and Deborah A. Deacon, *American Women Artists in Wartime, 1776–2010* (Jefferson, NC, and London, 2011)

Campany, David, 'Safety in Numbness: Some Remarks on the Problems of Late Photography', in *Where is the Photograph*, ed. David Green (Manchester, 2003), pp. 123–32

—, 'The Myth of Compassion Fatigue', in *The Violence of the Image: Photography and International Conflict*, ed. Liam Kennedy and Caitlin Patrick (London, 2014), pp. 97–124

Capa, Robert, *Slightly Out of Focus: Reminiscences of a War Photographer* (New York, 1947)

Carden-Coyne, Ana, *Reconstructing the Body: Classicism, Modernism, and the First World War* (Oxford, 2009)

Carmichael, Jane, *First World War Photographers* (London and New York, 1989)

Casasola Zapata, Gustavo, ed., *Historia gráfica de la Revolución Mexicana 1900–1970* (Mexico City, 1992)

Çelik, Zeynep, and Edhem Eldem, eds, *Camera Ottomana: Photography and Modernity in the Ottoman Empire, 1840–1914* (Istanbul, 2015)

Cloud, John, 'American Cartographic Transformations during the Cold War', *Cartography and Geographic Information Science*, XXIX/3 (2002), pp. 261–82

Collins, Kathleen, 'Living Skeletons: Carte-de-visite Propaganda in the American Civil War', *History of Photography*, XII/2 (1988), pp. 103–20

Coombs, Rose E. B., *Before Endeavours Fade: Guide to the Battlefields of the First World War*, 7th edn (London, 1994)

Cornell, Lauren, Julia Bryan-Wilson and Omar Kholeif, *Trevor Paglen* (New York and London, 2018)

Cuarterolo, Miguel Angel, 'Images of War', in *I Die with My Country: Perspectives on the Paraguayan War, 1864–1870*, ed. Hendrik Kraay and Thomas Whigham (London and Lincoln, NE, 2004), pp. 154–78

Daniel, Ute, Peter Gatrell, Oliver Janz, Heather Jones, Jennifer Keene, Alan Kramer and Bill Nasson, eds, *1914-1918-online: International Encyclopedia of the First World War* (Berlin, 2014)

Davis, Keith F., *George N. Barnard: Photographer of Sherman's Campaign* (Kansas City, MO, 1990)

Debroise, Olivier, *Mexican Suite: A History of Photography in Mexico*, trans. and rev. by Olivier Debroise and Stella de Sá Rego (Austin, TX, 2001)

Dewitz, Bodo von, *'So wird bei uns der Krieg geführt': Amateurfotografie im Ersten Weltkrieg* (Munich, 1989)

—, and Robert Lebeck, eds, *Kiosk: A History of Photojournalism, 1839–1973* (Cologne, 2001)

Diamond, Hanna, 'Representing Defeat: Photographic Images of the French Exodus of 1940', *Journal of War and Culture Studies*, I/3 (208), pp. 275–92

Didi-Huberman, Georges, *Images in Spite of All: Four Photographs from Auschwitz* (Chicago, IL, 2008)

Dorrian, Mark, 'Drone Semiosis: Weaponry and Witnessing', *Cabinet: A Quarterly of Art and Culture*, 54 (2014), pp. 48–55

Downing, Taylor, *Spies in the Sky: The Secret Battle for Aerial Intelligence during World War II* (London, 2011)

Duganne, Erina, 'Photography After the Fact', in *Beautiful Suffering: Photography and the Traffic in Pain*, ed. Mark Reinhardt, Holly Edwards and Erina Duganne (Chicago, IL, 2007), pp. 37–74

Earhart, David C., 'Kamikazefication and Japan's Wartime Ideology', *Critical Asian Studies*, xxxvii/4 (2005), pp. 569–96
Edwards, Elizabeth, *Raw Histories: Photographs, Anthropology and Museums* (Oxford, 2001)
Edwards, Holly, 'Cover to Cover: The Life Cycle of an Image in Contemporary Visual Culture', in *Beautiful Suffering: Photography and the Traffic in Pain*, ed. Mark Reinhardt, Holly Edwards and Erina Duganne (Chicago, IL, 2007), pp. 75–92
Eigen, Frauke, and Anna Catherina Gebbers, *Frauke Eigen: Fotografien 1999–2005* (Berlin, 2005)
Eisenman, Stephen F., *The Abu Ghraib Effect* (London, 2007)
Fehrenbech, Heidi, and Davide Rodogno, eds, *Humanitarian Photography: A History* (Cambridge, 2015)
—, and Davide Rodogno, '"A Horrific Photo of a Drowned Syrian Child": Humanitarian Photography and NGO Media Strategies in Historical Perspective', *International Review of the Red Cross*, xcvii/900 (2015), pp. 1121–55
Fernández, Horacio, ed., *Photobooks: Spain, 1905–1977* (Madrid, 2014)
Field, Ron, 'Camp Photographers: Pictures by the Thousand', *Military Images*, xxxv/4 (2017), pp. 60–64
Finnegan, Terrence, *Shooting the Front: Allied Aerial Reconnaissance in the First World War* (Washington, DC, 2006)
Ford, Colin, ed., *The Story of Popular Photography* (Bradford, 1989)
Fortuny, Natalia, 'Imágenes sobrevivientes: Fotografía y memoria en una obra de Paula Luttringer', *Clepsidra. Revista Interdisciplinaria de Estudios sobre Memoria*, 1 (2014), pp. 14–27
Friedrich, Ernst, *Krieg dem Kriege* (Berlin, *c.* 1930)
Friend, Melanie, *The Home Front* (Stockport, 2013)
Gallman, J. Matthew, and Gary W. Gallagher, eds, *Lens of War: Exploring Iconic Photographs of the Civil War* (Athens, GA, 2015)
Gamarnik, Cora, 'La construcción de la imagen de las Madres de Plaza de Mayo a través de la fotografía de prensa', *Afuera: Estudios de crítica cultural*, 9 (2010)
Gardner, Alexander, ed., *Gardner's Photographic Sketch Book of the War* (Washington, DC, 1866)
Germain, Éric, 'Out of Sight, Out of Reach: Moral Issues in the Globalization of the Battlefield', *International Review of the Red Cross*, xcvii/900 (2015), pp. 1065–97
Gervais, Thierry, 'The "Greatest of War Photographers": Jimmy Hare, A Photojournalist at the Turn of the Twentieth Century', trans. James Gussen, *Études photographiques*, http://journals.openedition.org, 26 November 2010
Godbey, Emily, '"Terrible Fascination": Civil War Stereographs of the Dead', *History of Photography*, xxxvi/3 (2012), pp. 265–74
Godbey, Michael, 'Confronting Horror: Emily Hobhouse and the Concentration Camp Photographs of the South African War', *Kronos*, xxxii (2006), pp. 34–48
Good, Jennifer, Paul Lowe, Brigitte Lardinois and Val Williams, eds, *Mythologizing the Vietnam War: Visual Culture and Mediated Memory* (Newcastle upon Tyne, 2014)
Gordon, Colin, ed., *Michel Foucault: Power/Knowledge: Selected Interviews and Other Writings* (London, 1980)
Gordon, Linda, and Gary Y. Okihiro, eds, *Impounded: Dorothea Lange and the Censored Images of Japanese American Internment* (New York, 2006)
Gordon, Sophie, *Shadows of War: Roger Fenton's Photographs of the Crimea, 1855* (London, 2017)
Gourevitch, Philip, and Errol Morris, *Standard Operating Procedure: A War Story* (London, 2008)
Gresh, Kristen, *She Who Tells a Story: Women Photographers from Iran and the Arab World* (Boston, MA, 2014)
Griffin, Michael, 'The Great War Photographs: Constructing Myths of History and Photojournalism', in *Picturing the Past: Media, History, and Photography*, ed. Bonnie Brennen and Hanno Hardt (Chicago and Urbana, IL, 1999)
—, 'Media Images of War', *Media, War and Conflict*, iii/7 (2010), pp. 7–41
Griffiths, Philip Jones, *Agent Orange: 'Collateral Damage' in Viet Nam* (London, 2003)
Guerin, Frances, *Through Amateur Eyes: Film and Photography in Nazi Germany* (Minneapolis, MN, 2012)
Halsall, Christine, *Women of Intelligence: Winning the Second World War with Air Photos* (Stroud, 2012)
Hamilton, Robert, 'Shooting from the Hip: Representations of the Photojournalist of the Vietnam War', *Oxford Art Journal*, ix/1, pp. 49–55
Hannavy, John, ed., *Encyclopedia of Nineteenth-century Photography* (New York and London, 2008)
Hardwood, Jeremy, *Hitler's War: World War II as Portrayed by Signal, the International Nazi Propaganda Magazine* (London, 2014)
Hare, James H., ed., *Collier's Photographic Record of the Russo-Japanese War* (New York, 1905)
Hariman, Robert, and John Louis Lucaites, *No Caption Needed: Iconic Photographs, Public Culture, and Liberal Democracy* (Chicago, IL, 2007)
Harris, David, *Of Battle and Beauty: Felice Beato's Photographs of China* (Santa Barbara, CA, 1999)
Hauser, Kitty, *Shadow Sites: Photography, Archaeology and the British Landscape, 1927–1955* (Oxford, 2007)
Heineman, Elizabeth D., ed., *Sexual Violence in Conflict Zones: From the Ancient World to the Era of Human Rights* (Philadelphia, PA, 2011)
Helbig, Elahe, 'Geographies Traced and Histories Told: The Photographic Documentation of Land and People by 'Abdullah Mirza Qajar, 1880s to 1890s', in *The Indigenous Lens? Early Photography in the Near and Middle East*, ed. Markus Ritter and Staci G. Scheiwiller (Berlin and Boston, MA, 2018), pp. 79–109
Herschdorfer, Nathalie, *Afterwards: Contemporary Photography Confronting the Past* (London, 2011)

Hesford, Wendy S., and Wendy Kozol, eds, *Haunting Violations: Feminist Criticism and the Crisis of the 'Real'* (Chicago and Urbana, IL, 2001)
Higonnet, Margaret Randulph et al., eds, *Behind the Lines: Gender and the Two World Wars* (New Haven, CT, 1987)
Hill, Jason, and Vanessa R. Schwartz, eds, *Getting the Picture: The Visual Culture of the News* (London and New York, 2015)
Hirsch, Marianne, *Family Frames: Photography, Narrative and Postmemory* (Cambridge, MA, 1997)
Hobhouse, Emily, *The Brunt of War and Where it Fell* (London, 1902)
Hodgson, Pat, *Early War Photographs* (Reading, 1974)
Hughes, Rachel, 'The Abject Artefacts of Memory: Photographs from Cambodia's Genocide', *Media, Culture and Society*, XXV (2003), pp. 23–44
Hüppauf, Bernd, 'The Emergence of Modern War Imagery in Early Photography', *History and Memory*, V/1 (1993), pp. 130–51
Issa, Rose, and Michket Krifa, eds, *Arab Photography Now* (Heidelberg and London, 2011)
Jenkins, R., ed., *Nagasaki Journey: The Photographs of Yosuke Yamahata: August 10, 1945* (Tokyo, 1995)
Kennedy, Liam, and Caitlin Patrick, eds, *The Violence of the Image: Photography and International Conflict* (London, 2014)
Kidd, Melanie, and Saleem Arif Quadry, eds, *Realism in Rawiya: Photographic Stories from the Middle East* (Nottingham, 2013)
Kriebel, Sabine T., *Revolutionary Beauty: The Radical Photomontages of John Heartfield* (Berkeley, CA, 2014)
Lee, Anthony W., and Elizabeth Young Berkeley, *On Alexander Gardner's Photographic Sketch Book of the Civil War* (Los Angeles, CA, 2007)
Leighton, George, 'The Photographic History of the Mexican Revolution', in *The Wind That Swept Mexico: The History of the Mexican Revolution 1910–1942*, ed. Anita Brenner (Austin, TX, 1984)
Lenman, Robin, ed., *The Oxford Companion to the Photograph* (Oxford, 2005)
Lewinski, Jorge, *The Camera at War: War Photography from 1848 to the Present Day* (London, 1978)
Life at War, with an Introduction by David E. Scherman (New York, 1981)
Long, J. J., Andrea Noble, and Edward Welch, eds, *Photography: Theoretical Snapshots* (Abingdon, 2009)
Longoni, Ana, 'Fotos y siluetas: dos estrategias en la representación de los desaparecidos', in *Los desaparecidos en la Argentina: Memorias, representaciones e ideas (1983–2008)*, ed. E. Crenzel (Buenos Aires, 2010), pp. 35–57
Longoni, Eduardo, *Imágenes apuntadas: Los relatos de un fotógrafo* (Buenos Aires, 2017)
Lorentzen, Lois Ann, and Jennifer Turpin, eds, *The Women and War Reader* (New York and London, 1998)
Lubben, Kristen, ed., *Susan Meiselas: In History* (New York, 2008)
Luther, Frederic, 'René Dagron, Inventor of Microfilm', *History of Photography*, XX/IV (1996), pp. 345–61
Luther, Narendra, *Raja Deen Dayal: Prince of Photographers* (Hyderabad, 2003)
Magubane, Peter, and Charlene Smith, *Soweto* (Cape Town, 2001)
Marcón, Barbara, 'Hiroshima and Nagasaki in the Eye of the Camera', *Third Text*, XXV/6 (2011), pp. 787–9
Marien, Mary Warner, *Photography: A Cultural History*, 3rd edn (London 2010)
Mathews, Chloe Dewe, *Shot at Dawn* (Madrid, 2014)
Meiselas, Susan, *Nicaragua: June 1978–July 1979* (New York, 1981)
—, and Mark Holborn, *Susan Meiselas: On the Frontline* (London and New York, 2017)
Menner, Simon, *Top Secret: Images from the Stasi Archives* (Berlin, 2013)
Moeller, Susan, *Compassion Fatigue: How the Media Sell Disease, Famine, War and Death* (New York, 1999)
—, *Shooting War: Photography and the American Experience of Combat* (New York, 1989)
Möller, Frank, *Visual Peace: Images, Spectatorship, and the Politics of Violence* (Basingstoke, 2013)
Moore, Alison M., 'History, Memory and Trauma in Photography of the Tondues: Visuality of the Vichy Past through the Silent Image of Women', *Gender and History*, XVII/3 (2005), pp. 657–81
Moutoussamy-Ashe, Jeanne, *Viewfinders: Black Women Photographers* (New York, 1986)
Mraz, John, *Photographing the Mexican Revolution: Commitments, Testimonies, Icons* (Austin, TX, 2012)
Murphy, Alistair, and Elizabeth Elmore, *Fishermen and Kings: The Photography of Olive Edis* (Norwich, 2016)
Nakhimovsky, Alexander, and Alice Nakhimovsky, *Witness to History: The Photographs of Yevgeny Khaldei* (New York, 1999)
Nelson, Patricia A., 'Competition and the Politics of War: The Global Photography Industry, *c.* 1910–60', *Journal of War and Culture Studies: The Business of War Photography*, IX/2 (2016), pp. 115–32
New Focus, eds, *No Man's Land: Young People Uncover Women's Viewpoints on the First World War* (Bradford, 2017)
Niven, Douglas, and Chris Riley, *The Killing Fields* (Santa Fe, NM, 1996)
Noble, Andrea, *Photography and Memory in Mexico: Icons of Revolution* (Manchester, 2010)
O'Gorman, Ned, and Kevin Hamilton, 'EG&G and the Deep Media of Timing, Firing, and Exposing', *Journal of War and Culture Studies: The Business of War Photography*, IX/2 (2016), pp. 182–201
O'Keefe, Daniel, ed., *Hurley at War: The Photography and Diaries of Frank Hurley in Two World Wars* (Sydney, 1986)
Oldfield, Pippa, ed., *Bringing the War Home* (Bradford, 2010)
—, 'Calling the Shots: Women's Photographic Engagement with War in the Hemispheric America, 1910–1990', PhD thesis, Durham University (2016)
—, 'Exploded Albums: Re-working the Family Photo Album in Women's Protests against State Terrorism in Argentina', in *The Photograph and the Album: Histories, Practices, Futures*,

ed. Jonathan Carson, Rosie Miller and Theresa Wilkie (Edinburgh and Boston, MA, 2013)
—, 'Snapshots from No Man's Land', www.britishartstudies.ac.uk, 2018
Özendes, Engin, *Photography in the Ottoman Empire, 1839–1919* (Istanbul, 1987)
Page, Tim, and William Shawcross, *Tim Page's Nam* (London, 1983)
Palmer, Kathleen, *Women War Artists* (London, 2011)
Palmquist, Peter E., *Elizabeth Fleischmann: Pioneer X-ray Photographer* (Berkeley, CA, 1990)
—, and Thomas R. Kailbourn, *Pioneer Photographers from the Mississippi to the Continental Divide: A Biographical Dictionary, 1839–1865* (Stanford, CA, 2005)
Penrose, Antony, ed., *Lee Miller's War: Photographer and Correspondent with the Allies, 1944–1945* (London, 1992)
Perlmutter, David D., *Photojournalism and Foreign Policy: Icons of Outrage in International Crises* (Westport, CT, 1998)
Peterson, Anne E., 'Alexander Gardner in Review', *History of Photography*, XXXIV/4 (2010), pp. 356–67
Pinney, Christopher, ed., *Photography's Other Histories* (Durham, NC, 2005)
Pino, Alberto del, 'Relaciones entre fotografía y demás iconografía de la Guerra del Paraguay', *Folia Histórica del Nordeste*, 25 (2016), pp. 133–58
Pritchard, Michael, *The History of Photography in 50 Cameras* (London, 2014)
—, *Spy Camera: A Century of Detective and Subminiature Cameras* (London, 1993)
Reinhardt, Mark, Holly Edwards and Erina Duganne, eds, *Beautiful Suffering: Photography and the Traffic in Pain* (Chicago, IL, 2007)
Reuter, Laurel, *Los Desaparecidos / The Disappeared* (Grand Forks, ND, 2006)
Roberts, Hilary, *Lee Miller: A Woman's War* (London and New York, 2015)
—, *Shaped by War: Photographs by Don McCullin* (London, 2010)
Roeder, George H., *The Censored War: American Visual Experience During World War Two* (New Haven, CT, 1995)
Romo, David Dorado, *Ringside Seat to the Revolution: An Underground Cultural History of El Paso and Juarez, 1893–1923* (El Paso, TX, 2005)
Rosenblum, Naomi, *A History of Women Photographers*, 3rd edn (New York, 2010)
Rosenheim, Jeff L., *Photography and the American Civil War* (New York, 2013)
Rosler, Martha, *Decoys and Disruptions: Selected Writings, 1975–2001* (Cambridge, MA, 2004)
Rubio, Oliva María, ed., *Margaret Bourke-White: Moments in History* (Madrid, 2013)
Samuels, Shirley, *Facing America: Iconography and the Civil War* (New York, 2004)
Schaber, Irme, Richard Whelan and Kristen Lubben, eds, *Gerda Taro: From the Collection of the International Center of Photography* (New York, 2007)
Shaul, Matthew, ed., *Beyond the Battlefields: Käthe Buchler's Photographs of Germany in the Great War* (Hatfield, 2018)
Sheehi, Stephen, *The Arab Imago: A Social History of Portrait Photography, 1860–1910* (Princeton, NJ, 2016)
Siddiqi, Asif, and Dwayne A. Day, 'Lies, Damned Lies, and Cover Stories: New Revelations About Satellite Secrecy During the Cold War', www.thespacereview.com, 20 November 2006
Sliwinski, Sharon, 'Visual Testimony: Lee Miller's Dachau', *Journal of Visual Culture*, IX (2010), pp. 389–408
Smith, Constance Babington, *Evidence in Camera: The Story of Photographic Intelligence in World War II* (London, 1957)
Smith, Lyn, *People Power: Fighting for Peace from the First World War to the Present* (London 2017)
Smith, Margaret Denton, and Mary Louise Tucker, *Photography in New Orleans: The Early Years, 1840–1865* (Baton Rouge, LA, and London, 1982)
Snow, Rachel, 'Photography's Second Front: Kodak's "Serving Human Progress Through Photography" Institutional Advertising Campaign', *Journal of War and Culture Studies: The Business of War Photography*, IX/2 (2016), pp. 151–81
Sontag, Susan, *Regarding the Pain of Others* (New York, 2003)
Spirit, Beverly A., and Patricia A. Randall, 'Radiologic History Exhibit: The Role of Women in Wartime Radiology', *Radiographics*, XV/3 (1995), pp. 641–52
Stallabrass, Julian, ed., *Memory of Fire: Images of War and the War of Images* (Brighton, 2013)
Stichelbaut, Birger, Jean Bourgeois, Nicholas Saunders and Piet Chielens, eds, *Images of Conflict: Military Aerial Photography and Archaeology* (Newcastle, 2009)
Stolarski, Christopher, 'The Rise of Photojournalism in Russia and the Soviet Union, 1900–1931', PhD thesis, John Hopkins University, Baltimore, MD (2013)
Struk, Janina, *Private Pictures: Soldiers' Inside View of War* (London, 2011)
Tada, Tsuguo, Miki Okabe and Juri Muramaki, eds, *Metamorphosis of Japan After the War* (London, 2011)
Taylor, Diana, *Disappearing Acts: Spectacles of Gender and Nationalism in Argentina's 'Dirty War'* (Durham, NC, 1997)
Taylor, John, *War Photography: Realism in the British Press* (London, 1991)
Tenfelde, Klaus, *Pictures of Krupp: Photography and History in the Industrial Age* (London, 2005)
Terry, Sara, ed., *War is Only Half the Story: 10 Years of the Aftermath Project* (Stockport, 2018)
Thomas, Adrian M. K., 'The First 50 Years of Military Radiology, 1895–1945', *European Journal of Radiology*, 63 (2007), pp. 214–19
Thomas, G., 'Indian Mutiny Veterans: The Tytlers', *History of Photography*, IX/4 (1985), pp. 267–73
Trachtenberg, Alan, 'Albums of War: On Reading Civil War Photographs', *Representations*, 9 (1985), pp. 1–32

Travnik, Juan, *Malvinas: Retratos y paisajes de guerra* (Buenos Aires, 2008)
Tucker, Anne Wilkes, and Will Michels, eds, *War/Photography: Images of Armed Conflict and its Aftermath* (New Haven, CT, 2012)
Tulloch, John, and R. Warwick Blood, 'Iconic Photojournalism and Absent Images: Democratization and Memories of Terror', in *The Routledge Companion to News and Journalism*, ed. Stuart Allan (Abingdon, 2010), pp. 507–19
Vanderwood, Paul J., and Frank N. Samponaro, *Border Fury: A Picture Postcard Record of Mexico's Revolution and U.S. War Preparedness, 1910–1917* (Albuquerque, NM, 1988)
Verduyn, Christl, and Conny Steenman Marcusse, 'Emmy Andriesse, Dutch Wartime Photographer: The Hunger Winter of 1945', in *Bearing Witness: Perspectives on War and Peace from the Arts and Humanities*, ed. Sherrill Grace, Patrick Imbert and Tiffany Johnstone (Montreal, 2012), pp. 121–44
Versteeg, Wilco, 'War Beyond Photography: Digitally Embellished Imaging and Contemporary Conflicts', *Arts of War and Peace Review*, II/1 (May 2014), pp. 33–51
Vettel-Becker, Patricia, *Shooting from the Hip: Photography, Masculinity, and Postwar America* (Minneapolis, MN, 2005)
Villela, Samuel F., *Sara Castrejón: Fotógrafa de la Revolución* (Mexico City, 2010)
Virilio, Paul, *War and Cinema: The Logistics of Perception* (London, 1989)
Vis, F., and O. Goriunova, eds, *The Iconic Image on Social Media: A Rapid Research Response to the Death of Aylan Kurdi* (Sheffield, 2015)
Wallace, A. F., 'The Early History of Clinical Photography for Burns, Plastic and Reconstructive Surgery', *British Journal of Plastic Surgery*, XXXVIII (1985), pp. 451–65
Wang, Caroline, and Mary Ann Burris, 'Photovoice: Concept, Methodology, and Use for Participatory Needs Assessment', *Health Education and Behaviour*, XXIV/3 (1997), pp. 369–87
Weld, Kristen, *Paper Cadavers: The Archives of Dictatorship in Guatemala* (Durham, NC, 2014)
Wellington, Jennifer, *Exhibiting War: The Great War, Museums, and Memory in Britain, Canada, and Australia* (Cambridge, 2017)
Wenger, Gina L., 'Documentary Photography: Three Photographers' Standpoints on the Japanese-American Internment', *Art Education*, LX/5 (2007), pp. 33–8
West, Nancy Martha, *Kodak and the Lens of Nostalgia* (Charlottesville, VA, 2000)
Whelan, Richard, and Cornell Capa, eds, *Robert Capa: The Definitive Collection* (London, 2004)
White, William, 'The Microdot: Then and Now', *International Journal of Intelligence and Counter Intelligence*, III/2 (1989), pp. 249–69
Williams, Val, *Warworks: Women, Photography and the Iconography of War* (London, 1994)
Wilson, Jacqueline Z., Sarah Hodgkinson, Justin Piché and Kevin Walby, eds, *The Palgrave Handbook of Prison Tourism* (London, 2017)
Zelizer, Barbie, 'Finding Aids to the Past: Bearing Personal Witness to Traumatic Public Events', *Media, Culture, and Society*, XXIV/5 (2002), pp. 697–714
—, *Remembering to Forget: Holocaust Memory through the Camera's Eye* (Chicago, IL, 1998)
Zeller, Bob, *The Blue and Gray in Black and White: A History of Civil War Photography* (Westport, CT, 2005)
—, *The Civil War in Depth: History in 3-D* (San Francisco, CA, 1997)
Zervigón, Andrés Mario, *John Heartfield and the Agitated Image: Photography, Persuasion, and the Rise of the Avant-garde Photomontage* (Chicago, 2012)
Zerwes, Erika, 'The Mother from Extremadura and the Idea of a Photographic Icon', *Roots and Routes: Research on Visual Cultures*, V/17 (2015)

Acknowledgements

My sincere thanks to the team at Reaktion, especially Vivian Constantinopolous, Editorial Director, for her invaluable expertise, insight and professionalism, and to Phoebe Colley for shepherding the book to completion. I am indebted to series editor Peter Hamilton for taking the project on, and for his guidance and unstinting support throughout: it has been a pleasure to work with you. I am also indebted to the numerous 'critical friends' who commented on the draft: Tom Allbeson, Geoffrey Belknap, Pete Brook, Erina Duganne, Melanie Friend, Dominic Hopkinson, Wendy Kozol, J. J. Long, Anne McNeill, John Oxley, Tony Penrose, Martha Rosler, Sophie Wright and especially to Robin Lenman.

I am most grateful to the British Academy for a BA/Leverhulme Small Research Grant, which made this publication possible. My thanks also to Durham University, Paul Mellon Centre for Studies in British Art, and Peter E. Palmquist Memorial Fund for Historical Photographic Research, for grants that supported the research that helped shape this book.

Of the numerous experts who generously shared their time and knowledge, I would particularly like to acknowledge the scholars Petra Bopp, Eldem Edem, Elahe Helbig, Alberto del Pino Menck and Samuel Villela; as well as Alison Metcalfe at National Library of Scotland, and Ruth Neave at BAPRAS. I am immensely grateful to the individual photographers, artists and collectors for allowing me to reproduce images in their possession, and to Mae Whitlock Gentry and Leonardo Mellibovsky for sharing their family histories.

I would like to thank those individuals who, over the years, helped me develop my thinking on the intersection of war and photography, including the photographers Melanie Friend, Lisa Barnard and Alison Baskerville, and IWM curator Hilary Roberts. Thanks as ever to my colleagues at Impressions Gallery, in particular Anne McNeill, Jen Sobol and the members of New Focus, and all the artists who contributed to the exhibitions Bringing the War Home and No Man's Land: Women's Photography and the First World War.

I learned a huge amount from contributors to The Business of War Photography: Producing and Consuming Images of Conflict, a conference I co-convened with Tom Allbeson at Durham University; and from colleagues at workshops in which I participated, particularly Cold War Camera in Guatemala and Visualising Peace at Edinburgh University. As an alumna and Honorary Research Fellow at Durham University, I am especially indebted to my colleagues at the School of Modern Languages and Cultures, and to my former PhD supervisors for their guidance, encouragement and generosity: Jonathan Long, Ed Welch, and – above all – the much-missed Andrea Noble (1968–2017).

Photo Acknowledgements

The author and publishers wish to express their thanks to the following sources of illustrative material and/or permission to reproduce it:

ADAGP, Paris and DACS, London 2019: p. 71; After the Battle Publications, Old Harlow, Essex: p. 173; © the artist (Jananne Al-Ani) and Abraaj Capital Art Prize: p. 72; Album House, Golestan Palace, Tehran: p. 56; Anti-Kriegs-Museum, Berlin: p. 152; AP (The Associated Press)/ Shutterstock: p. 37; Art Gallery of New South Wales, Sydney: p. 146; © Bandan Atta/Metrography: p. 54; Archive of Auschwitz-Birkenau State Museum, Oświęcim: p. 140; © the artist (Lisa Barnard): p. 88; © the artist (Alison Baskerville/Impressions Gallery): p. 48; the British Association of Plastic, Reconstructive and Aesthetic Surgeons, London (Gillies Collection): p. 87; British Library, London, © British Library Board, all rights reserved/Bridgeman Images: p. 46; Brown University Library (Brown Digital Repository, Anne S. K. Brown Military Collection), Providence, RI: p. 162 (left); © the artist (Tessa Bunney): p. 188; © Robert Capa © International Center of Photography © Magnum Photos © 1937 Time Inc. All rights reserved. Reprinted from LIFE and published with permission of Time Inc. Reproduction in any manner in any language in whole or in part without written permission is prohibited: p. 29; © Consuelo Castrejón, courtesy Samuel Villela: p. 95; Nigel J. Clarke/ Alamy Stock Photo: p. 66; © the artist (Jill Cole): p. 105; Crown Copyright: pp. 67 (courtesy of the Medmenham Collection), 103 (courtesy of the Council of the National Army Museum, London (NAM. 1998-10-244-1); © the artist (Gohar Dashti): p. 183; © Nilüfer Demir/AFP/Getty Images/ with kind permission of Liz Sly: p. 115; Deutsches Museum, Munich (Archive, CD51942): p. 81; © the artist (Chloe Dewe Mathews)/Ruskin School of Art at University of Oxford/14–18 NOW: p. 181; Digital Globe Inc., courtesy Amnesty International: p. 78; Drik Picture Library, Dhaka, Bangladesh: p. 53; David C. Earhart Collection of Japanese Primary Sources from the Asia-Pacific War: p. 99; reproduced with permission of Eastman Kodak Company (collection of the author): p. 76; © the artist (Frauke Eigen): p. 179; © the artist (Hasan Elahi): p. 127; courtesy Edhem Eldem Collection: p. 128; © the artist (Léa Eouzan): p. 174; Feminist Archive at University of Bristol Library (Special Collections): p. 120; © the artist (Melanie Friend): p. 171; Galerija 11/7/95, Sarajevo: p. 169; courtesy Mae Whitlock Gentry: p. 162 (right); courtesy of the George Eastman Museum, Rochester, NY: p. 166; photo by Ronald S. Haeberle/ The LIFE Images Collection/Getty Images: p. 147; Harvard Law School Library (Historical and Special Collections), Cambridge, MA: p. 158; © The Heartfield Community of Heirs/DACS 2019: p. 117; © Imperial War Museum, London: pp. 43 (Q 11745), 64 (Q 47823), 93 (Q 30038), 101 (Documents.2031); © Philip Jones Griffiths/Magnum Photos: p. 178; JR Inside Out Project (© the artists: Ali Rez, Saks Afridi, Assam Khalid, Akash Goel, Insiya Syed, Noor Behram, Jamil Akhtar): p. 73; photo by Mustafa Kamaci/Anadolu Agency/Getty Images: p. 39; Doug Kanter/ AFP/Getty Images: p. 108; © the artists (kennardphillipps): p. 121; Leiden University Libraries (Special Collections): p. 139; Library of Congress, Washington, DC (Prints and Photographs Division): pp. 18, 23, 26, 27, 35, 60, 62, 111, 114 (© The Regents of the University of California, The Bancroft Library, University of California, Berkeley), 143 and 163 (Liljenquist Family Collection of Civil War Photographs); © the artist (Eduardo Longoni): p. 155; © the artist (Paula Luttringer): p. 137; © Don McCullin/Contact Press Images: p. 107; © the artist (Joe McNally): p. 185; Madres de Plaza de Mayo Linea Fundadora, Buenos Aires, with permission of Leonardo Mellibovsky: p. 156; © Juliet Madsen/ Veterans Books Project: p. 186; Manzanar War Relocation Center Records, Library Special Collections, Charles E. Young Research Library, UCLA, Los Angeles, CA: p. 136; © Susan Meiselas/Magnum Photos: p. 182; courtesy of the artist (Simon Menner)/BSTU (Bundesbeauftragte für die Stasi-Unterlagen), Berlin: p. 126; © the artist (Lana Mesic): p. 190; Metropolitan Museum of Art, New York: pp. 10, 34; © Lee Miller Archives, Chiddingly, East Sussex, 2019, with permission of Condé Nast Publications Ltd: p. 144; © Wayne Miller/Magnum Photos: p13; courtesy of Massachusetts Institute of Technology Museum, Cambridge, MA (© 2010 MIT): p. 82; Museo Histórico Nacional, Montevideo, Uruguay/ courtesy Alberto del Pino: p. 50; Archivo fotografico del Museo Nacional Centro de Arte Reina Sofia, Madrid – photograph by David Seymour/ Magnum Photos: p. 112; Museum für Photographie Braunschweig (Deposit, Stadtarchiv Braunschweig – Estate of Käthe Buchler): p. 94; © 2019 The Museum of Modern Art/Scala, Florence/Chris Riley, Photo Archive Group: p. 133; © Nabiela 2002/PhotoVoice/Aschiana/ 'Shooting Kabul'/Afghanistan: p. 187; National Archives and Records Administration, College Park, MD (records of the War Relocation Authority): p. 135; National Archives and Records Administration, Northeast Region New York City (Records of Naval Districts and Shore Establishments): p. 98; National Library of Medicine, Bethesda, MD: p. 85; © National Library of Scotland: p. 44; Collection of the National Museum of African American History and Culture, Smithsonian Institution, Washington, DC (Gift of Gina R. McVey, granddaughter): p. 52; National Security Archive, George Washington University, Washington, DC (Dino A. Brugioni Collection): p. 69; National Science and Media Museum, Bradford/Science and Society Picture Library: p. 176; collection of the author: pp. 41, 100, 110; courtesy of the artist (Trevor Paglen), © Trevor Paglen, Metro Pictures New York, Altman Siegel San Francisco: p. 74; Pump Park Vintage Photography/Alamy Stock Photo: p. 51; © the artist (Wendy Red Star): p. 167; © George Rodger/ Magnum Photos/Library of Congress, Washington, DC (Prints and Photographs Division, New York World-Telegram and Sun Collection): p. 5; Roger-Viollet/TopFoto: p. 123; © the artist (Martha Rosler): p. 118; © RMN-Grand Palais (Musée d'Orsay, Paris)/photo Hervé Lewandowski: p. 109; reproduced by permission of the Royal Engineers Museum,

Library and Archive, Gillingham, Kent: p. 59; © the artist (Seivan M. Salim): p. 149; Sovfoto/UIG via Getty Images: p. 32; Stasimuseum, Berlin/ ASTAK e.V./Photo: John Steer: p. 125; State Library of New South Wales, Sydney: p. 31; © Swiss Camera Museum, Vevey: p. 79; © Gerda Taro, Collection International Center of Photography (90.2005): p. 47; © Newsha Tavakolian/Magnum Photos: p. 164; © the artist (Juan Travnik): p. 177; © the artist (Jonathan Turner): p. 170; University of Leeds Library Special Collections (Liddle Collection): p. 160; War Museum of the Boer Republics, Bloemfontein, South Africa: p. 151; Wellcome Collection, London: p. 86; Collection of Dale West: p. 22; photo by Alex Wong/Getty Images: p. 159; courtesy of Shogo Yamahata: p. 145.

Index

Illustration numbers are in *italic*